The Urban Book Series

Aims and Scope

The Urban Book Series is a resource for urban studies and geography research worldwide. It provides a unique and innovative resource for the latest developments in the field, nurturing a comprehensive and encompassing publication venue for urban studies, urban geography, planning and regional development.

The series publishes peer-reviewed volumes related to urbanization, sustainability, urban environments, sustainable urbanism, governance, globalization, urban and sustainable development, spatial and area studies, urban management, urban infrastructure, urban dynamics, green cities and urban landscapes. It also invites research which documents urbanization processes and urban dynamics on a national, regional and local level, welcoming case studies, as well as comparative and applied research.

The series will appeal to urbanists, geographers, planners, engineers, architects, policy makers, and to all of those interested in a wide-ranging overview of contemporary urban studies and innovations in the field. It accepts monographs, edited volumes and textbooks.

More information about this series at http://www.springer.com/series/14773

Matthias Finger · Maxime Audouin
Editors

The Governance of Smart Transportation Systems

Towards New Organizational Structures for the Development of Shared, Automated, Electric and Integrated Mobility

Springer

Editors
Matthias Finger
Chair Management of Network Industries
College Management of Technology
Lausanne, Switzerland

Maxime Audouin
Chair Management of Network Industries
College Management of Technology
Lausanne, Switzerland

ISSN 2365-757X ISSN 2365-7588 (electronic)
The Urban Book Series
ISBN 978-3-030-07208-7 ISBN 978-3-319-96526-0 (eBook)
https://doi.org/10.1007/978-3-319-96526-0

This Springer imprint is published by the registered company Springer Nature Switzerland AG
The registered company address is: Gewerbestrasse 11, 6330 Cham, Switzerland

Contents

Editors and Contributors

About the Editors

Matthias Finger received his Ph.D. degrees in political science and adult education from the University of Geneva, Switzerland. He is known for his expertise in matters of regulation and governance of network industries (postal services, telecommunications, electricity, railways, air transportation, urban public transportation, water). He has been an assistant professor at Syracuse University (New York), an associate professor at Columbia University (New York), and a full professor at the Swiss Federal Institute of Public Administration (Lausanne). Since 2002, he has held the Swiss Post Chair in Management of Network Industries (MIR) at Ecole Polytechnique Federale de Lausanne (EPFL), Lausanne, Switzerland; and since 2010, he has directed the Florence School of Regulation's Transport Area at the European University Institute, Florence, Italy.

Maxime Audouin holds a Bachelor of Science in Environmental Engineering, and a Master of Science in Energy Management, both from Ecole Polytechnique Federale de Lausanne (EPFL), Switzerland. Since 2015, he has been enrolled as a Ph.D. student in the Swiss Post Chair in Management of Network Industries (MIR) at EPFL, under the supervision of Prof. Matthias Finger. Audouin's research interests lies in the area of urban transportation systems, urban governance, transportation policy and regulation, as well as technology and innovation management. In his dissertation, Audouin looks at the role that public authorities play in the development of ICT-supported integrated mobility schemes.

Contributors

Maxime Audouin Chair Management of Network Industries (MIR), College Management of Technology (CDM), EPFL, Lausanne, Switzerland

Bruno Faivre d'Arcier Transport Urban Planning Economics Laboratory, University of Lyon, Lyon, France

Matthias Finger Chair Management of Network Industries (MIR), College Management of Technology (CDM), EPFL, Lausanne, Switzerland

I. C. MariAnne Karlsson Chalmers University of Technology, Gothenburg, Sweden

Yveline Lecler Institute of East Asian Studies, University of Lyon, Sciences Po Lyon, Lyon, France

Yanying Li European Climate Foundation, Brussels, Belgium

Theo Lieven Institute for Customer Insight, University of St. Gallen, St. Gallen, Switzerland

Henk Meurs Department of Transport and Planning, Faculty of Civil Engineering and Geosciences, Delft University of Technology, Delft, The Netherlands

Miloš N. Mladenović Spatial Planning and Transportation Engineering, Department of Built Environment, Aalto University, Espoo, Finland

Juan J. Montero UNED University, Madrid, Spain; European University Institute, Florence School of Regulation, Florence, Italy

Dalia Mukhtar-Landgren K2—The Swedish Knowledge Centre for Public Transport, Lund, Sweden; Lund University, Lund, Sweden

Maria Lorena Puche Independent Researcher, Miami, USA

Nele Rietmann Institute for Customer Insight, University of St. Gallen, St. Gallen, Switzerland

Marianne Ryghaug Department of Interdisciplinary Studies of Culture, Norwegian University of Science and Technology (NTNU), Trondheim, Norway

Steven Sarasini RISE Viktoria, Gothenburg, Sweden

Fariya Sharmeen Department of Transport and Planning, Faculty of Civil Engineering and Geosciences, Delft University of Technology, Delft, The Netherlands; Department of Geography, Planning and Environment, Institute for Management Research, Radboud University, Nijmegen, The Netherlands

Tomas Moe Skjølsvold Department of Interdisciplinary Studies of Culture, Norwegian University of Science and Technology (NTNU), Trondheim, Norway

Göran Smith Chalmers University of Technology, Gothenburg, Sweden; K2—The Swedish Knowledge Centre for Public Transport, Lund, Sweden; Västra Götalandsregionen, Gothenburg, Sweden

Jana Sochor Chalmers University of Technology, Gothenburg, Sweden; RISE Viktoria, Gothenburg, Sweden

Tom Voege ITF-OECD, Paris, France

Chapter 1
Introduction

Maxime Audouin and Matthias Finger

Abstract Mainly because of the dominance of private automobiles, most transportation systems are known to be ill-functioning nowadays, especially in urban environments. However, thanks to recent technological developments, new transport solutions appear as formidable opportunities to break away from the incumbent automobility regime and move toward more sustainable transport systems. Yet, mainly because of their highly disruptive nature and their provision by private and new actors, the integration of those new mobility options in existing transportation systems calls for the development of new organizational structures. In this chapter, we provide an overview of how smart transportation systems are conceptualized, by focusing on their four foundational pillars—they are shared, automated, electric and have integrated mobility—and explain why these pillars call for the development of new governance approaches. We conclude by highlighting the research gap that we aim to tackle and by briefly presenting the different contributions that are gathered in this book.

Keywords Governance · Shared mobility · Electric mobility · Automated mobility · Integrated mobility

1.1 Introduction

One could consider that transportation systems are to cities what cardiovascular networks are to human bodies (Samaniego and Moses 2008). In theory, they are supposed to seamlessly connect the different subparts of urban systems so those

---

M. Audouin (✉) · M. Finger
Chair Management of Network Industries (MIR), College Management of Technology (CDM), EPFL, Lausanne, Switzerland
e-mail: maxime.audouin@epfl.ch

M. Finger
e-mail: matthias.finger@epfl.ch

M. Finger and M. Audouin (eds.), *The Governance of Smart Transportation Systems*, The Urban Book Series, https://doi.org/10.1007/978-3-319-96526-0_1

interact interdependently with one another, ultimately structuring cities into gigantic living organisms (Hommels 2005). In reality, however, urban transportation systems do not flow anywhere near as smoothly as blood in most people's blood vessels. Indeed, mainly because of the predominance of automobiles on roads, which are known to have become counter-productive (Illich 1973), cities are very often congested and ultimately synonymous with ill-functioning urban systems. What the diseased cities, and more generally contemporary societies, are suffering from is often referred to as automobility, which is defined as a complex path-dependent nonlinear system centered on fossil-fuel-powered private cars and supported by a range of institutions and infrastructures (Urry 2004). Being described as a "modern Leviathan" or a "Frankenstein-created monster" (Latimer and Munro 2006; Urry 2004), automobility is usually considered responsible for massive space and environmental resource consumption, congestion, as well as noise pollution and casualties resulting from road traffic accidents. For example, the annual cost of congestion (most of which occurs in urban areas) in Europe is approximately 1% of the EU's GDP (130 billion euros). Similarly, two-thirds of the 750,000 road accidents that occur each year in Europe take place in cities, and in 2015, these led to 9700 fatalities (EU 2017). Thus, given that urban populations and mobility demands of individuals are increasing globally (ITF 2017), the need to unlatch cities from the automobility lock-in has never appeared to be more urgent than it is today.

French philosopher Roland Barthes described the automobile as the gothic cathedral of modern times, that is, as a man's made creation that individuals would deeply worship, having necessitated the most advanced state of technological and scientific knowledge to come to life, and filled with symbolism (Barthes 1957). However, as was the case with cathedrals in contemporary society, the place privately owned petrol automobiles occupy in transportation systems might fade away with time. According to Urry (2004), there are several so-called turning points that might enable to transit toward the "post-car" system. Among those, new fuel systems, information and communication technologies (ICTs), smart cards, and new consumption patterns relying on usership rather than traditional ownership might pave the way toward a renewed transportation system that is less auto destructive and more sustainable than the incumbent one.

In recent decades, the developments in transportation have mainly occurred along the four above-mentioned turning points. As we will detail below, the mobility industry has more specifically undergone four major disruptions, with the development of shared mobility (relying on new consumption patterns), automated mobility (relying on ICTs), electric mobility (relying on new propulsion technologies), as well as integrated mobility (relying, among other things, on smart cards) solutions.

1.2 The Four Pillars of Smart Transportation Systems

The first disruption, and first pillar of smart transportation systems, is referred to as the *shared mobility* pillar. Advances in ICTs have enabled significant developments in geo-localization systems, which are increasingly embedded in smartphones and have paved the way for the development of new transport options (Shaheen et al. 2009), relying on the sharing of a specific asset (which can be a vehicle) or of a dedicated service (a ride, for example), that is to say new consumption patterns. Shared mobility services providers have developed, particularly in cities, offering consumers the ability to achieve automobility while avoiding the burden of vehicle ownership and its associated with sunk fixed costs (Shaheen et al. 2018). Although such solutions have existed for quite some time, it is acknowledged that the ICTs have enabled those to takeoff by considerably reducing associated transaction costs and greatly simplifying access to it. Shared mobility services can actually be divided into four distinct groups: carsharing, car-pooling, e-hailing, and demand-responsive transit (DRT). Carsharing (sometimes referred to in the UK as car clubs) refers to the sharing of a vehicle (or pool of vehicles) between individuals. Carsharing can actually take four different forms being round-trip, one way, P2P carsharing, or fractional ownership (Shaheen et al. 2018). Carsharing services are usually organized by dedicated companies that propose digital platforms enabling users to search for a car, book it, and pay for it online. Car-pooling refers to the sharing of a vehicle journey by the vehicle driver, so that other people can join the ride, which ultimately enables the driver to share the costs of his or her trip. Car-pooling services are usually also organized via dedicated online platforms, where drivers can propose their trips and passengers can search for a trip that matches their needs (or at least partly) and book and pay for the service online. Thirdly, e-hailing services (also sometimes referred to as ride-sharing, ride-sourcing, or ride-booking) refer to the booking of a private ride via a dedicated online platform, usually established by a transportation network company (TNC) that enables the pairing of drivers (amateurs and professionals) with passengers. The main difference with car-pooling is that, in the case of e-hailing, the driver provides the ride based on the passenger's needs and does so for money, whereas in car-pooling it is the driver who chooses the route based on his or her needs, and decides to welcome other passengers if it fits theirs. Car-pooling drivers usually do not make detours and it is up to the passengers to accept the driver's conditions, not the other way around. Last but not least is demand-responsive transit (also sometimes referred to as ride-pooling), which refers to a demand-based semi-public transit system. As opposed to having fixed transit lines operating on fixed routes on a fixed schedule, DRT operates when and where users demand it, which they usually do thanks to a smartphone app or Web interface. DRT vehicles are usually minibuses that hold between six- and fourteen-seated passengers whose routes are determined by a ride-pooling algorithm in charge of finding the shortest route to access the different passengers' destinations. Overall, shared mobility solutions enable a shift from automobile ownership to mobility usership (Shaheen et al. 2018) and shared mobility service providers often present their solutions as having the ability to reduce road traffic, congestion, and pollution (Cohen and Kietzmann 2014).

The second disruption, and second pillar of smart transportation systems, is referred to as the *automated mobility* pillar. Within the ICT revolution, advances in sensing and communication technologies have enabled the development of so-called self-driving vehicles (SDVs), which are able to sense their environment through dedicated sensors, analyze the data gathered, and make decisions accordingly in order to navigate with reduced or no human input. Different levels of vehicle automation exist ranging from limited assistance to full automation. Proponents of SDVs usually advocate that most traffic accidents are the result of human error, and that enabling machines to take care of the driving tasks could significantly reduce the number of vehicle crashes and transport-related casualties, and ultimately increase the efficiency of the overall road transport system (Fagnant and Kockelman 2015). In this book, the terms automated vehicle (AV) and self-driving vehicle (SDV) are used interchangeably.

The third disruption, and what we refer to as the third pillar of smart transportation systems, pertains to *electric mobility*. Despite not having been enabled directly by the ICTs, electric mobility represents a whole array of new solutions that rely on new propulsion technologies and are supposed to emit much less greenhouse gases and other pollutants than conventional internal combustion engines (ICE) vehicles (Orsato 2017). Although electric vehicles have existed for almost a century, it is only in the last few decades that their share of total vehicles on roads has started to increase significantly. In this book, the notion of electric vehicles basically encompasses the notions of battery electric vehicles (BEVs), hybrid electric vehicles (HEVs), and plug-in hybrid electric vehicles (PHEVs), all of which are different. BEVs use electric motors instead of traditional ICE for propulsion and use chemical energy stored in dedicated battery packs to run their electric motors. HEVs are equipped with both a conventional ICE and an electric propulsion system, aiming to achieve a better performance in terms of fuel consumption than conventional ICE vehicles. PHEVs are a subcategory of HEVs, which can basically receive an external electric power source to charge their electric batteries. In this case, the combustion engine acts as a backup when electric batteries are empty. Although the increase in electric mobility has been seen as the direct result of increased sales in EVs, it has also been enabled thanks to the increased use of EVs by shared mobility services providers, particularly by car clubs operating in urban environments.

While some have argued that mobility is “only” currently undergoing a threefold disruption, with developments in automated, electric, and shared mobility solutions (Sperling 2018), we believe that transportation systems are also developing along a fourth and last axis. This ultimate pillar of smart transportation systems, and fourth disruption of traditional transport systems, is referred to as *integrated mobility*, which aims to enable multimodal travel—defined as the combination of two or more different transport services, public or private, to accomplish a journey—and produce a shift from private motorized travel to more sustainable modes of travel such as public transport or shared mobility modes. Integrated mobility has basically been facilitated recently by two main ICT-supported developments: the development of integrated multimodal information systems, and integrated payment solutions. While the former has enabled users to access and compare specific travel information in real time from different transport providers, and therefore pick the solution best fitting their mobility needs (Kenyon and Lyons 2003), the latter has enabled users to access various transportation solutions with a single ticketing means, which could be a card (smart card) or a dedicated app (Morfoulaki et al. 2015). Put together, and also supported by the birth of new shared mobility solutions, those two ICT-supported developments have enabled the unfolding of the Mobility-as-a-Service (MaaS) concept, defined as a digitally supported distribution model bundling several transport options together, and enabling users to plan their trips, select the transport option that best suits their needs, and finally book and pay for it via an app. Given its potential ability to help breaking away with car ownership and eventually pave the way for more sustainable transport systems and cities, MaaS has been triggering a lot of interest from transport scholarship lately.

While those new disruptive transport seems to be bringing some of the changes needed to exit the automobility regime, they also call for the development of new governance structures to ensure they do not actually pave the way for another path-dependent and autopoietic transportation regime.

1.3 The Need for New Governance Structures

It is still unclear how the traditional transport world is reacting to the introduction of smart transport solutions that private actors are usually proposing. Those new solutions are indeed disrupting the status quo and calling for the development of new organizational approaches. For example, Transportation Network Companies have completely disrupted the conventional taxi business, pushing public authorities to react and ultimately decide whether to embrace or ban e-hailing services, for which urban citizens are usually quite supportive. Although SDVs hold the promise of making roads safer and road transport more efficient, they also call for the development of new rules, as automated vehicles will basically be sharing streets, boulevards and avenues with conventional human-driven vehicles as well as pedestrians—who we know do not think as “perfectly” as computers. Similarly, the development of MaaS solutions by the private sector is pushing traditional transport actors (operators and authorities) to re-invent themselves, and decide quickly what part they want to play in this new game, with the threat of being finally eaten alive if they refuse to play. One might also wonder how governments are reacting to the unfolding of electric mobility, which holds the promise of a partly decarbonized transport system, and how they are, or are not, institutionally framing their development. While most of the actors of the four above-mentioned disruptive trends impacting mobility advocate the ability of their solutions to help break away from the incumbent automobility regime, their true ability to produce a sustainable impact on transport systems actually depends on the way they unfold (Docherty 2018).

Although there is not a single and simple answer to the question of how smart transport solutions are being integrated in existing transportation systems, we argue that the response lies in the field of transport governance. Despite the existence of a wide range of definitions, governance is used in this book in a quite general sense to define "*the means by which to infuse order, thereby to mitigate conflict and realize mutual gain*" (Williamson 2005: 43). Although it has not been used extensively in transport research, scholars have increasingly emphasized the importance of the governance concept in the organization of sustainable (urban) transportation systems (e.g. Banister 2008; Hull 2008; Schiller et al. 2010). Also, as traditional transportation systems are undergoing a fourfold transformation, it appeared natural to look at this disruption from the governance perspective. While most research to date on smart transportation has been concerned with understanding the impacts of those disruptive options on existing transportation systems (see, for example, Meyer and Shaheen 2017), very little research (recent exceptions include Marsden and Reardon 2018) has been carried out about the governance of smart transportation systems. Therefore, we hope the present book, by bundling together a series of case studies looking at the underlying processes that have supported (or hindered) developments in either shared, automated, electric, and integrated mobility solutions, contributes to tackling this gap. In particular, the book covers two aspects of smart transportation systems governance. On the one hand, some chapters will be looking at regulatory governance of smart transport solutions development. On the other hand, contributions will be adopting a more socio-technical approach to transportation and will focus more on the governance of transition toward the post-car system, as we detail below.

1.4 Outline of the Book

As explained above, smart transportation systems can be understood as being composed of four pillars (shared, automated, electric, and integrated). Accordingly, we have structured the book into four main sections, each of which focuses on one of the above-mentioned pillars.

The first section looks at the governance of shared mobility solutions. The first chapter in this section, by Juan Montero, is entitled "*Regulating Transport Platforms: The Case of Car-pooling in Europe*" and uses literature on multi-sided markets to discuss the nuts and bolts of many novel-shared mobility solutions, using car-pooling in Europe as a case study. Montero argues that as most shared mobility solutions are actually provided through dedicated platforms, public authorities must envisage new regulatory approaches, particularly the development of a level playing field with traditional players.

The second chapter of this section, by Maria Lorena Puche, entitled "*Regulating TNCs in Latin America: The Case of UBER regulation in Mexico City and Bogota*," explores how governments in Latin America have been reacting, from a regulatory standpoint, to the entry of e-hailing services in their cities. Through a case study approach exploring the development of e-hailing regulation in Colombia and Mexico, Puche shows that the use of traditional taxi regulations by public authorities to regulate TNCs is not always effective, and that new and more innovative regulatory frameworks need to be developed.

The third chapter of this section, by Bruno Faivre d'Arcier and Yveline Lecler, entitled "*Governing Carsharing as a Commercial or a Public Service? A Comparison between France and Japan,*" focuses on the institutional conditions having supported the development of one way and round-trip carsharing in Japan and France. The authors conclude that, mainly due to different positions of public authorities vis-à-vis this shared mobility solution, the development of carsharing has followed opposite paths in both countries. While France has mainly seen one way carsharing come to life, primarily due to its development as a public service, it is mainly commercial round-trip carsharing services that have managed to sustain in Japan.

The second part of the book looks at the governance of automated mobility, which stands as the second pillar of smart transportation systems, as described earlier. The fifth chapter of the book (and first of this second section), by Tom Voege, is entitled "*Data-led Governance of Self-Driving Vehicles for Urban Shared Mobility.*" Voege discusses the link between the first and second pillars of smart transportation systems, that is, between shared and automated mobility. According to Voege, as TNCs will inevitably start proposing automated e-hailing services, public authorities will need to be a step ahead and develop data-driven regulations that are able to cope with this future disruption, and ultimately kill two birds with one stone.

The second chapter of this section (and sixth chapter in the book) is by Miloš N. Mladenović and entitled "*How Should We Drive Self-Driving Vehicles? Anticipation and Collective Imagination in Planning Mobility Futures.*" Building on science and technology studies, the chapter discusses the importance of understanding technology as a socio-technical phenomenon to guarantee successful development of new transport technologies. Using the case of SDVs deployment in Finland, Mladenović proposes eventually replacing the language of unintended consequences with the language of unanticipated consequences, as well as developing a phase of participatory expansion when new (transport) technologies are introduced, to plan for (un)desirable futures.

The third part of the book focuses on the governance of electric mobility, which is the third major trend disrupting transportation systems. This section begins with the seventh chapter of the book, by Nele Rietmann and Theo Lieven, entitled "*A Comparison of Policy Measures Promoting Electric Vehicles in 20 Countries,*" which aims to compare the impact of EVs supportive policy measures on EVs market share. Results indicate that the higher the level and amount of political incentives in a country, the higher the proportion of EVs in the country. The two authors also conduct a qualitative analysis on two of the 20 countries they analyze (the Netherlands and Brazil) to illustrate that governance structures, and not only policy measures, are important in the uptake of electric vehicles by individuals.

The eighth chapter of the book is by Marianne Ryghaug and Tomas Moe Skjølsvold and is entitled "*Nurturing a Regime Shift Toward Electro-mobility in Norway.*" Using the multi-level perspective on socio-technical transition framework, the authors explain how electric mobility has developed in Norway. While they agree that policy measures have played a role in the uptake of electric vehicles, they also demonstrate that, in the case of Norway, those policies did not target primarily the decarbonization of transport, but were actually defined to stimulate the development of a Norwegian EV industry, illustrating not only how policies work, but why and how they are coming to life.

The fourth and final section of the book looks at the governance of integrated mobility. It starts with the chapter by Göran Smith, Steven Sarasini, I.C. MariAnne Karlsson, Dalia Mukhtar-Landgren and Jana Sochor, entitled "*Governing Mobility-as-a-Service: Insights from Sweden and Finland.*" Using transition management, the authors look at the role that Finnish and Swedish public authorities at the local and national level are playing in the development of MaaS schemes. One of their conclusions is that in order for MaaS schemes to develop in a sustainable way, public actors must align their operational and tactical activities related to the development of MaaS with other relevant transport strategies.

The second chapter of this fourth section (and the tenth chapter of the book), by Maxime Audouin and Matthias Finger, is entitled "*Institutions, Organizations, and Technological Innovations: The Dynamic Development of Smart Ticketing Schemes in London's Urban Transportation System.*" Using a conceptual framework building on systems of innovation literature and co-evolution between technology and institutions theory, the authors explore the case of smart card development in London. Through their case study, they highlight the fact that while it is evident that institutions influence how organizations interact with one another, and how this leads to the emergence of technological innovations, the influence that technological innovations have on organizations, which in turn might shape new institutional developments, must not be ignored.

The eleventh chapter of the book, by Fariya Sharmeen and Henk Meurs, is entitled "*The Governance of Demand-Responsive Transit Systems—A Multi-level Perspective.*" The authors basically establish the link between the last and the first sections of the book by examining the development of DRT as a preliminary step in the development of MaaS schemes. Looking at a case of DRT development in the Netherlands, and using the multi-level perspective framework, the authors explore the barriers and drivers for the diffusion of demand-responsive transit. They conclude that while the willingness of the local government and transit operators can be understood as drivers in DRT developments, infrastructure, technology, and market practices are more likely to act as barriers.

In the last chapter of the last section, entitled "*The Role of Public Authorities in the Development of Mobility-as-a-Service (MaaS),*" Yanying Li investigates, through a cross-case analysis, the different approaches of public authorities vis-à-vis the development of MaaS schemes (GoDenver in Denver, UbiGo in Sweden, Whim in Helsinki, and WienMobil in Vienna). Li highlights that the output of MaaS schemes will ultimately depend largely on the way they unfold, and some MaaS solutions might not be as sustainable as they might pretend. Li concludes by recommending that, in order for MaaS to maximize public good, public authorities must urgently define clearer KPIs to know where they wish to go.

Maxime Audouin and Matthias Finger conclude the book in the thirteenth chapter by summarizing the key themes covered in the book. They reflect on their notions of smart transportation solutions and recall the need to conciliate the "smart" perspective with the sustainability perspective. The editors finish by making recommendations on future research related to the development of smart transportation systems.

We hope all those chapters, taken together, offer a comprehensive overview of the different research endeavor focusing on the governance of smart transportation systems and act as a stepping stone for more research related to this trendy and important subject, crucial for the future of cities.

References

Banister D (2008) The sustainable mobility paradigm. Transp Policy 15(2):73–80

Barthes R (1957) Mythologies. Seuil, Paris

Cohen B, Kietzmann J (2014) Ride on! Mobility business models for the sharing economy. Organ Environ 27(3):279–296

Docherty I (2018) New governance in the era of smart mobility. In: Marsden G, Reardon L (eds) Governance of the smart mobility transition. Emerald Points

European Commission (2017) European mobility, policy context. DG MOVE, European Commmission, Brussels

Fagnant D, Kockelman K (2015) Preparing a nation for autonomous vehicles: opportunities, barriers and policy recommendations. Transp Res Part A 77:167–181

Hommels A (2005) Studying obduracy in the city: toward a productive fusion between technology studies and urban studies. Sci Technol Human Values 30(3):323–351

Hull A (2008) Policy integration: what will it take to achieve more sustainable transport solutions in cities? Transp Policy 15(2):94–103

Illich I (1973) Energie et Equité. Seuil, Paris

ITF (2017) ITF transport outlook 2017. OECD and International Transport Forum, Paris

Kenyon S, Lyons G (2003) The value of integrated multimodal traveller information and its potential contribution to modal change. Transp Res Part F Traffic Psychol Behav 6(1):1–21

Latimer J, Munro R (2006) Driving the social. In: Böhm S, Jones C, Land C, Paterson M (eds) Against automobility. Blackwell Sociological Review Monograph, Oxford

Marsden G, Reardon L (2018) Governance of the smart mobility transition. Emerald Points

Meyer G, Shaheen S (2017) Disrupting mobility: impacts of sharing economy and innovative transportation on cities. Springer, Berlin

Morfoulaki M, Myrovali G, Kotoula K (2015) Increasing the attractiveness of public transport by investing in soft ICT based measures: going from words to actions under an austerity backdrop—Thessaloniki's case, Greece. Res Transp Econ 51:40–48

Orsato R (2017) Future imperfect: the enduring struggle for electric vehicles. In: Nieuwenhuis P, Vergragt P, Wells P (eds) The business of sustainable mobility: from vision to reality. Routledge, New York, pp 35–44

Samaniego H, Moses ME (2008) Cities as organisms: allometric scaling of urban road networks. J Transp Land Use 1(1):21–39

Schiller PL, Bruun EC, Kenworthy JR (2010) An introduction to sustainable transportation: policy, planning and implementation. Earthscan/James & James, Oxford
Shaheen S, Cohen A, Chung M (2009) North American carsharing: 10-year retrospective. Transp Res Rec 2110:35–44
Shaheen S, Martin E, Bansal A (2018) Peer-to-peer (P2P) carsharing: understanding early markets, social dynamics, and behavioral impacts. UC Berkeley Research Report. https://doi.org/10.7922/G2FN14BD
Sperling D (2018) Three revolutions: steering automated, shared, and electric vehicles to a better future. Island Press, Washington
Urry J (2004) The "system" of automobility. Theory Cult Soc 21(4/5):25–39
Williamson O (2005) Transaction cost economics. In: Menard S (ed) Handbook of new institutional economics. Springer, Berlin

Maxime Audouin holds a Bachelor of Science in Environmental Engineering, and a Master of Science in Energy Management, both from Ecole Polytechnique Federale de Lausanne (EPFL), Switzerland. Since 2015, he has been enrolled as a Ph.D. Student in the Swiss Post Chair in Management of Network Industries (MIR) at EPFL, under the supervision of Prof. Matthias Finger. Audouin's research interest lies in the area of urban transportation systems, urban governance, transportation policy and regulation, as well as technology and innovation management. In his dissertation, Audouin looks at the role that public authorities play in the development of ICT-supported integrated mobility schemes.

Matthias Finger received his Ph.D. degrees in Political Science and Adult Education from the University of Geneva, Switzerland. He is known for his expertise in matters of regulation and governance of network industries (postal services, telecommunications, electricity, railways, air transportation, urban public transportation, water). He has been an assistant professor at Syracuse University (New York), an Associate Professor at Columbia University (New York), and a Full Professor at the Swiss Federal Institute of Public Administration (Lausanne). Since 2002 he has held the Swiss Post Chair in Management of Network Industries (MIR) at Ecole Polytechnique Federale de Lausanne (EPFL), Lausanne, Switzerland, and since 2010 has directed the Florence School of Regulation's Transport Area at the European University Institute in Florence, Italy.

Part I
Governance of Shared Mobility

Chapter 2
Regulating Transport Platforms: The Case of Carpooling in Europe

Juan J. Montero

Abstract Online platforms are transforming transportation as they create new multi-sided markets connecting transport providers (often non-professional providers) and passengers, in a way that generates new network effects and distributes them among the participants in the ecosystem. The European carpooling experience shows the power of transport platforms to multiply the traditional sharing of resources and even to substitute collective transportation modes such as railways and buses. Transport platforms raise new regulatory challenges. New regulations must consider that the efficiency created by the platforms derives from pooling together large volumes of users. This is possible if platforms are allowed to provide their intermediation services without taking all the burden of a transport service provider. However, a balanced relationship between the platforms and the service providers on the one side, and the passengers on the other side, has to be ensured. Furthermore, a level playing field with traditional players has to be ensured. Finally, the potential market power has to be taken into consideration.

Keywords Sharing economy · Carpooling · Platform · Regulation BlaBlaCar

2.1 Introduction

In recent years, transport has been disrupted by the emergence of online platforms mediating between new transport service providers, often non-professional service providers, and passengers. Transport platforms like Uber, Lyft, Didi, and Ola are transforming urban mobility all around the globe. BlaBlaCar is transforming long-distance traveling, particularly in Europe.

J. J. Montero (✉)
UNED University, Madrid, Spain
e-mail: jlmontero@der.uned.es

J. J. Montero
European University Institute, Florence School of Regulation, Florence, Italy

M. Finger and M. Audouin (eds.), *The Governance of Smart Transportation Systems*, The Urban Book Series, https://doi.org/10.1007/978-3-319-96526-0_2

Disruption by online platforms is not specific to transport. Such disruption is one of the most obvious effects of digitalization across industries. It started in the music and media industries, but it has expanded to telecoms, energy, and finance (Montero and Finger 2018). Transport platforms are just another example of the power of digitalization to transform economic activity.

Digitalization takes the form of the creation of a data layer on top of the physical world. Sensors extract data from the physical world (from computers, from smartphones, from Internet of Things sensors, etc.) and a parallel virtual map of the world is constructed. Artificial intelligence (AI) makes it possible to automate the management of the large amount of data extracted from reality. New and creative interactions are made possible. In particular, new network effects can be created in the data layer.

Platforms are at the center of this transformation. They concentrate the data extracted from the physical word, they have Artificial Intelligence capabilities to manage such data and to extract value from it, and they are in the position to create and curate new multi-sided markets and to make new network effects possible.

Platforms bring efficiency to the transport industry. Firstly, they reduce transaction costs, empowering new service providers, even non-professional service providers. Secondly, platforms allow individual, isolated, service providers to act as a coordinated network. A network of individual vehicles is far more efficient than isolated, uncoordinated, vehicles. Finally, platforms can act as coordinators of the system at a larger scale, creating a network with all the existing transport service providers across transport modes, both for urban and long-distance transport.

However, platforms pose very significant regulatory challenges: (1) The role of platforms from a legal perspective; (2) a level playing field has to be ensured as old and new transport modes are now competing; (3) transport service providers, particularly new non-professional service providers, might find themselves under undesired contract conditions ("uberization"); (4) passengers' rights might be under challenge in the new model; and (5) platforms are gaining market power, perhaps to an excessive point.

Such challenges can only be met if regulation adapts to the new industrial organization model. Old solutions that ignore the role of a platform in a multi-sided market will not solve the new challenges. In the worst-case scenario, efficiencies created by platforms will be destroyed, harming consumers and damaging competitiveness in a global economy.

This chapter aims to provide guidance on the new regulatory challenges created by transport platforms, using carpooling as the study case. Carpooling can be defined as the shared use of a vehicle by a driver (usually the owner of the vehicle) and one or more passengers in order to divide the cost of a trip made fully or partially together. In carpooling, the driver makes the trip with the passenger. This is different from other sharing schemes in which only the vehicle is shared, without the presence of the driver/owner. (Such a solution is usually named carsharing and the leading company is Car2Go). In carpooling, the driver is making the trip out of his or her own interest to move from one point to another. This contrasts with other models, where the driver is merely providing a transport service to the passenger. Such a solution is usually named e-hailing, and it has been championed by Uber.

Section 2.2 describes the new platform paradigm and the economic theory developed around it. Section 2.3 describes the historical evolution of carpooling from the historic origins to the success of carpooling platforms in Europe. Section 2.4 analyzes in some detail how carpooling platforms work and the basis for its success. Section 2.5 builds on the previous analysis to identify the main regulatory challenges posed by transport platforms and provides some guidance on how to meet them. Section 2.6 provides some conclusions.

2.2 Online Platforms

2.2.1 Platforms in Multi-sided Markets

The seminal work of Rochet and Tirole (2003) entitled "Platform Competition in Two-Sided Markets" has spawned a substantial amount of industrial organization literature: Evans and Schmalensee (2016), Evans (2011), Libert et al. (2016), Parker et al. (2016). However, there is no commonly agreed definition of online platform or of "two-sided" or "multi-sided market." Such a difficulty was identified at an early stage by Rochet and Tirole, who warned about a "*you know a two-sided market when you see it*" approach (Rochet and Tirole 2006: 645).

Over the years, multi-sided markets have been shown to typically involve two or more distinct types of users, interacting through a third party—the platform—which acts as an intermediary (OECD 2011). Key elements are the relevance of positive indirect network effects and the leading role of the platform in the fair distribution of such benefits across the ecosystem of interacting parties, through pricing and other design decisions.

A typical example of multi-sided market is newspapers and the media in general. A platform—the newspaper—allows the interaction between readers and advertisers. The key to the success of a newspaper is indirect network effects. The wider the number of users in each side, the higher the benefit for the other side. On the one side, advertisers will be increasingly interested in the newspaper as it is more widely read. On the other side, readers will be increasingly interested in a newspaper if content is enriched with the revenue generated by a larger pool of advertisers. The platform, the newspaper, has a significant role as it defines the distribution of the benefits derived from the new interaction among all the players: the readers, the advertisers, and the platform itself. If the benefit in terms of higher advertising revenue is not shared with the readers in terms of higher expenditure in quality content (and a lower price for the newspaper), readers will not buy the newspaper. Newspapers are an example of non-transactional multi-sided markets,[1] as the different parties interacting through the platform do not conclude transactions among themselves.

Payment cards are an example of transactional multi-sided markets as the platform—the payment card company—facilitates direct transactions between merchants and purchasers. The larger the pool of merchants accepting the card, the more attractive the card is for purchasers. The more purchasers use the card, the more attractive it is for merchants. The market grows because the larger the number of users on both sides, the greater the benefits for all of them.

Other examples of traditional platforms in multi-sided markets are stock exchanges (connecting investors and companies looking for investors), commercial centers (connecting retailers and shoppers), and dating agencies.

2.2.2 *Online Platforms*

The Internet has multiplied the power of platforms in multi-sided markets. As shown above, platforms in multi-sided markets are an old mode of industrial organization, even though they have only recently received specific attention from the economic literature. It was the introduction of the Internet that really multiplied the interest of academics, practitioners, and investors in this form of industrial organization.

Most Internet champions are actually platforms in multi-sided markets. Google is a platform in the search market that connects advertisers, content producers, and individuals looking for content; Facebook is a platform in the social network market that connects advertisers and social media users; Apple is a platform in the app market that connects mobile app developers and app users; Skype is a platform in the communications market that connects users in different telecom networks; and Airbnb is a platform in the accommodation market that connects hosts and guests.

The key role of the Internet in enhancing platforms is not surprising. The very nature of the Internet is precisely to facilitate interconnection, namely the interaction of previously fragmented and isolated telecommunications systems. As communications are made universal (Internet empowers anyone to communicate with anyone), ubiquitous (thanks to mobile networks), and inexpensive, new and creative forms of interaction among individuals emerge.

As the Internet reduces transaction costs, producers and consumers can interact directly, thus reducing the role of traditional intermediaries (record companies, travel agents, real estate agents, etc.). However, intermediaries are not really abolished; rather, traditional intermediaries are replaced by new intermediaries, namely the online platforms. The growing role of intermediaries in the Internet era

[1]On the differences between transactional and non-transactional markets, see Damme et al. (2010).

might be counterintuitive. However, it is explained by the fact that those intermediaries that are disappearing are traditional intermediaries in traditional one-sided markets: mere "middlemen." Instead, the intermediaries that are now proliferating are the ones that create and structure multi-sided markets. Online platforms, as intermediaries in multi-sided markets, are necessary in order to create the conditions for the different sides to interact. They provide the framework for communication (the apps). They create the algorithms that automatically match specific individuals on each side of the market. They "curate" the contracting conditions, particularly the pricing structure, but also the payment mechanism, the trust mechanisms (evaluations, "stars," "likes," etc.). The role of the new intermediaries is to create and curate the market so that the different sides can interact efficiently.

Successful online platforms rely on two distinct but mutually reinforcing factors. Firstly, online platforms, just like the traditional platforms, reduce transaction costs and therefore provide an attractive alternative for the different sides to interact. Inexpensive communications, automatized matching, trust-building instruments, and the definition of an efficient pricing structure are all elements that reduce transactions costs.

Secondly, the reduction of transaction costs increases scale in the platforms, making indirect network effects possible. As platforms attract a large pool in each side of the market, a positive externality is created for the other side. The more passengers use a transport platform, the more drivers will be interested in joining the platform; vice versa, the more drivers are available, the more passengers will be interested in joining the platform. The Internet offers the possibility of creating multi-sided markets where literally billions of individuals interact at a global scale.

The benefits derived from the indirect network effects, if fairly distributed among the different sides active in the market (and the platform itself), reduce the cost of the provision of the service, bringing new demand to the market. A virtuous cycle is created that "ignites" (grows the large volumes of users on all sides of the market) the multi-sided market and the platform structuring it. This is the successful business model that is also disrupting transportation.

2.3 Carpooling: From Its Origins to Success in Europe

2.3.1 Origins and Evolution

Carpooling has a long history. Informal sharing of private motor vehicles probably started as soon as the first motor vehicles were available, in the late nineteenth century. Formal programs to share trips were introduced in the USA in 1942, as part of the war effort, to reduce consumption of scarce resources (gasoline, rubber, etc.). Factories, churches, and parent-teacher associations were asked to form "carsharing club exchanges" (Chan and Shaheen 2012).

As a response to the energy crisis of the 1970s, similar initiatives spread around the USA (Pratsch 1975; Weiner 1999), and Europe (ATEMA/ADEME 2010). These initiatives usually focused on commuting carpooling rather than long-distance carpooling.

As environmental awareness increased in the late 1990s, carpooling programs became popular again. The Internet became the tool to connect drivers and passengers. Carpooling was mostly restricted to specific communities (companies, universities, etc.) as members of a community have common transportation needs and the necessary trust among members. So-called "casual carpooling"—drivers picking-up passengers in designated areas to reach the minimum number of passengers to use reserved lanes—reached relevance in only a handful of US metropolitan areas, led by San Francisco (Kelly 2007). In any case, carpooling was mostly restricted to urban commutes and to a small scale.

2.3.2 Sharing Economy and Online Platforms

It was only after 2010 that carpooling reached a substantial volume. Two factors are immediately related to this evolution: the sharing economy and online platforms.

Sharing, defined as using an asset jointly, either at the same time or in turns, is as old as humanity. However, over the last decade, a new socioeconomic model has emerged around the notion of sharing both assets and services. It is the "sharing economy" (Sundararajan 2016), also called "collaborative consumption" (Botsman and Rogers 2010).

The following factors are fueling the new model. Firstly, citizens, organizations, and public authorities are increasingly aware of the negative environmental effects of lifestyles in industrialized societies. Transport represents almost a quarter of Europe's greenhouse gas emissions and is the main cause of air pollution in cities (European Commission 2016: 2). Secondly, increasing economic hardship is generating a new demand for low-cost services. Air transportation is a good example, as is the proliferation of low-cost bus services (Steer Davies Gleave 2016). Thirdly, there is a general trend toward choosing access over ownership. Individual ownership of assets is being replaced by the possibility to use assets without owning them. Consumers are increasingly aware of the cost of owning assets in terms of maintenance, repairs, insurance, storage, etc., as well as of the externalities in the form of congestion, environmental damage, and so on. The reduction in the rate of private vehicle ownership in the most developed societies is a good example, as is the even more significant trend among young people of either not obtaining a driver's license or putting off taking their license exam (Beck 2016).

Online platforms are playing a leading role in the sharing economy (Montero and Finger 2018). Online platforms concentrate the technological innovations for the interaction of the parties interested in sharing goods and services. Parties interested in sharing identify themselves on the platform choose the goods and services to be shared, their location, the conditions, and so on. Platforms match the owners of assets to be shared with users interested in making use of them. Algorithms automatize the matching process (McAfee and Brynjolfsson 2017), which takes place at a very low cost. The concentration in the platform of very large pools of providers and consumers creates the fundamental indirect network effects that ignite a platform.

Trust is made possible thanks to the new cultural values facilitated by the social networks, as well as by specific instruments provided by the platform such as identification of the users, evaluations, "likes," and artificial intelligence tools that help to manage ratings and exclude fraudulent ratings. Trust among users of online platform is decisive. Carpooling platforms have achieved a high level of trust among users, often even higher than trust in colleagues and neighbors.[2] Trust is particularly necessary in carpooling, as passengers have to rely on the driving ability of an unknown person and risk is high.

Lower transaction costs multiply the possibilities to share (Munger 2015). Sharing is mostly possible when the transaction cost is lower than the value obtained by the individuals involved in a sharing transaction. Traditionally, sharing was limited to transactions with low transaction costs (sharing with relatives or friends), or to high-value transactions. Now that transaction costs are very low, sharing is possible even when the value of the transaction is very low. It is possible to share, just for a few hours, an under-used asset such as a hand-drill, and along these lines, it is possible to share a bicycle, a car—or a seat in a car.

All these trends crystallized by the end of 2008 and the beginning of 2009, when the leading sharing economy platforms were established: Airbnb was created in San Francisco in November 2008, Uber in San Francisco in March 2009, BlaBlaCar was set up in Paris in 2006, but it was only in 2009 when the app for mobile smartphones was launched that the platform ignited. Transportation has been one of the leading areas of the sharing economy from its inception.

2.3.3 *The Ignition of Carpooling Platforms*

Long-distance carpooling through online platforms has been particularly popular in Europe and it is mostly a European success story. BlaBlaCar has 40 million members in 22 countries, mostly in Europe. It claims to match 12 million trips per quarter and to provide access to 2 million trips at any given moment (BlaBlaCar 2017).

---

[2]Based on a survey of 18,289 BlaBlaCar users in 11 countries, Mazzela and Sundarajan (2016) stated that 88% of respondents had a very high or high trust in a member with a full profile in the platform. The same level of trust in colleagues was declared by only 58% of respondents, and in neighbors by 42% of participants of the survey. Trust in family reached 94%.

BlaBlaCar is the market leader for carpooling in all the large European countries, with markets' shares above 90% in France, Spain, Germany, Italy, and Eastern Europe.[3]

Carpooling exists in the USA, but is far less popular than in Europe, particularly for long-distance trips. In the USA, companies like Zimride and rdvouz.com are active in the long-distance carpooling market, but far from the volumes of users and mediated rides of BlaBlaCar in Europe. BlaBlaCar has launched operations in Brazil, Mexico, and India, but the volume of these operations has not yet reached the levels of success of the European operations.

Carpooling has reached a significant portion of the overall long-distance transportation market in the more mature European markets (Finger et al. 2017). According to CGDD (2016), a total of 8 million carpooling trips were made in France in 2015, for a total of 6 billion passenger kilometers (pkm).[4] Those numbers are summed-up in Tables 2.1 and 2.2.

To put this into perspective, 6 billion pkm represent 2.72% of the total pkm in domestic long-distance transportation in France in 2015, and more than 12% of the pkm of long-distance railway transportation.[5] In economic terms, it can be estimated that the fees paid by carpooling passengers to drivers in France, in 2015, amounted to EUR 210 million.[6]

CGDD (2016) estimates that carpooling has a significant potential for growth, even in the most mature markets. Carpooling in France could grow from 6 billion to 14 billion pkm.

Table 2.1 Carpooling in France 2015

Item	Estimate
Number of trips offered	8 million
Number of passengers transported	11 million
Average distance by trip	320 km
Number of passenger-km (not including driver)	3.5 million
Number of passengers (including driver)	19 million
Number of passengers-km (including driver)	6 billion

Adapted from CGDD (2016)

[3]BlaBlaCar grew in Germany through the acquisition in April 2015 of its local competitor, carpooling.com, which had 6 million registered users. BlaBlaCar grew in Eastern Europe through the acquisition in March 2015 of Budapest-based AutoHop and in January 2016 of Jizdomat, active in the Czech Republic and Slovakia.

[4]This figure is coherent with the figures in other studies. According to UIC (2016), the number of carpooling trips in 2015 was estimated at 7 million (p. 55). Based on public statements by BlaBlaCar, GART/UTP (2014) estimated the number of passengers transported in 2013 to be around 1% of the total number of long-distance passengers and 5% of the trips in public land transportation modes (train, bus, and carpooling).

[5]In France, trips of more than 80 km are considered long distance.

[6]This is the result of multiplying the 3.5 billion pkm by the average price per km (EUR 0.06).

Table 2.2 Long-distance passenger-km per mode in France in 2015

Mode	Passenger-km (in billion)	%
Private car	158	71.98
Of which total carpooling users	6	2.72
Of which carpooling passengers	3.5	1.59
Of which carpooling drivers	2.5	1.13
Train	49	22.32
Airplane	13	5.92
All	219.5	100

Adapted from CGDD (2016)

Carpooling is growing around Europe. The increasing relevance of carpooling in terms of number of users, number of pkm, and payments that have been quantified for France can serve as a lesson for the not-yet-mature (in terms of carpooling presence) markets in the rest of Europe.

2.4 Factors Enhancing Carpooling in Europe

2.4.1 Indirect Network Effects

The key to the success of carpooling in Europe has been the accumulation of very significant indirect network effects as large pools of drivers and passengers joined that platform.

Large pools of drivers and passengers are necessary to ensure the availability of rides for passengers, and passengers for drivers. The larger the volumes of drivers, the greater the chances of finding a driver making the desired trip at the right time. The larger the number of passengers, the greater the chances that a driver will find one or more passengers.

At the early stage of BlaBlaCar, only trips between large urban areas were available in the platform, and such trips tended to be concentrated along weekends and holidays. As the pool of users became larger, trips between large metropolitan areas and smaller towns became more common. Finally, as the number of registered users reached the millions, trips even between smaller towns became available.

The same evolution can be identified for times of departure. Convenience is important. According to ADEME/6T (2015), the time of departure is the main factor when choosing one driver over another (89%), even more than the price (79%) or the type of vehicle (22%). Shaheen et al. (2016) identified saving time as one of the main reasons passengers use the service. However, availability of the service diverges very significantly among routes. Dense routes among large cities ensure a wide range of options, often more options than collective transportation (night services, strikes, cross-border services). Along these lines, GART/UTP

(2014) identified that carpooling is stronger when the railway service is weaker. Routes connecting rural areas, on the contrary, provide fewer options or even no option at all.

As the number of passengers increases, carpooling becomes more attractive. On denser routes, the occupation rate of a car is higher than on other routes (CGDD 2016: 6). The higher the occupation rate of a car, the lower the price of the service, as costs can be distributed among a larger pool of passengers.

Since low prices and flexibility in departure times are the key parameters to use carpooling, the service improves as the number of users increases, generating a further increase in the number of users. Such a virtuous cycle is the defining trait of the indirect network effect.

2.4.2 *Distribution of the Benefit Derived from Internalized Network Effects*

A factor for the success of carpooling in Europe is that BlaBlaCar identified the right equilibrium in the distribution of the indirect network effects created by the platform, in terms of pricing, both for drivers and passengers, and commissions for the platform.

Economic reasons are the main driver for the growth of long-distance carpooling: 69% of carpooling users chose the service for this reason, 12% for the flexibility in the schedule, 7% for the duration of the trip, and 7% for the social experience (ADEME/6T 2015). This finding is consistent across all surveys. Carpooling prices are clearly below railway prices (particularly high-speed services) and are usually below bus service prices (see Sect. 2.5.2).

Since price is the driver of carpooling usage, it is of interest to identify how prices are set. The leading carpooling platform, BlaBlaCar, recommends that drivers set the price by dividing the cost of gasoline and tolls by three. Prices rarely diverge substantially from this reference. In France, it has been identified that the average price per passenger is EUR 0.06 per km (CGDD 2016: 5).

There are several reasons for the low prices of carpooling services. Firstly, drivers tend to share only variable costs (gasoline and tolls) with passengers, while fixed costs (cost of the vehicle, insurance maintenance etc.), which amount to two-thirds of the total cost in France (GART/UTP 2014: 42), are often ignored. This could be because the driver is traveling to the destination anyway, and not merely transporting third parties to the destination. Secondly, national regulation on carpooling often introduces a limit on fees that can be charged, as the provision of the service for a profit is often prohibited. Thirdly, compared to traditional collective transportation services, carpooling has some regulatory advantages; for example, the driver does not pay income taxes when charging passengers, no time limitations or public service obligations are imposed on drivers, some countries do not charge tolls for the use of roads, and the main cost borne by railway undertakings is the access charge for the use of the railway infrastructure.

Therefore, even if private vehicles are not more efficient than railways or buses, they can charge lower prices to passengers, thus affecting passengers' choice of transport mode.

The leading online platform, BlaBlaCar, charges passengers a commission of around 18%. It is interesting to emphasize that the commission is charged to the passenger and not to the driver; this is contrary to the charging model of other transport platforms, which charge a similar commission, but to drivers.

Economic reasons are also the most relevant for drivers to share their rides. According to surveys among carpooling drivers in France, 29% declared that they saved more than EUR 50 in their last trip, 27% saved between EUR 20 and 50, and 39% less than EUR 20 (ADEME/6T 2015). The fact that fees received by drivers do not seem particularly high could explain why carpooling does not attract a large share of wealthy drivers, but rather young professionals with salaries below the national average (see Sect. 2.4.3).

In conclusion, carpooling is successful in Europe as fees are low enough for passengers to compete with other transport modes, while high enough for drivers to go the trouble of sharing their drives.

2.4.3 Migration of Passengers from Long-Distance Mass-Transit to Carpooling

Existing data strongly suggests that the success of carpooling in Europe is explained by the migration of a substantial number of passengers from traditional long-distance collective transport services to carpooling (Finger et al. 2017). The existence of a large number of passengers of collective services seems to be an important factor in the success of long-distance carpooling.

In France and the rest of Europe, millennials were the early adopters that helped to grow the carpooling service, and they remain the main users. BlaBlaCar users in France are younger than the average population. Passengers (average age of 34) tend to be younger than drivers (average 37), when the average age of all people in France is 40 (ADEME/6T 2015). In any case, the average age of the users is growing as the service matures. It is particularly interesting that the fastest growing group in 2015 was that of users above 60, another group with a low rate of private car use. Carpooling users tend to be better educated than the average[7] and tend to be single and without children (Oudghiri and Brunet 2013).

Surveys have identified that long-distance carpooling passengers in France have a low rate of car ownership (ADEME/6T 2015: 44). Only 52% of leisure carpooling passengers own a car. This is a very low rate compared with the overall population. Low ownership rates are particularly acute among very young passengers. This can

[7]Forty-seven percent of users have a college degree, compared to 13% of the total population (ADEME/6T 2015). Similar results can be found in Shaheen et al. (2016).

be explained by the fact that passengers are often students and employees with low salaries. Drivers, tend to be young professionals with salaries below the average (Shaheen et al. 2016).

It has been identified that carpooling is having an impact on car ownership. According to ADEME/6T (2015), 3% of carpooling users declared that they had given up owning a car, and 11% of users declared that they delayed acquiring a car either because they postponed the procurement of a driver's license (7%) or the acquisition of the car itself (4%). As passengers have more alternatives, owning a car becomes an option that can be delayed or dismissed.

In this framework, the large volume of pkm in carpooling in Europe is due mostly to the migration of passengers from collective services to carpooling. It is not the result of the shift from the use of private vehicle to carpooling. In fact, most carpooling passengers and even some carpooling drivers were previously traveling by train or bus. The following evidence confirms the analysis.

Firstly, the academic literature and some empirical research underline the strong competition between carpooling and mass-transit transportation, both for commuting and for long-distance trips (Minett and Pearce 2011; TCRP 2012; Godillon 2016).

Secondly, various studies have shown that carpooling is competitive with collective services in terms of both price and duration of the trip. A study by UIC (2016) included comparative tables on the substitutability of the different long-distance transport modes, including carpooling (see Table 2.3). The data shows that carpooling is a good substitute for railway and bus services.

Table 2.3 Paris–Lille: modal chain, total time, and cost

Mode of transport	Transfer and waiting time from origin to main mean of transport	Duration of trip-leg with main mean of transport	Transfer and waiting time from main mean of transport to final destination	Total travel time from origin to destination	Total cost	Time	Money
Rail	55 min	1 h	35 min	2h30	40€	++	–
Bus	55 min	3 h	35 min	4h20	18€	–	+
Car	7.5 min	2 h	7.5 min	2h15	40€	++	–
Carpooling (as driver)	22.5 min	2 h	22.5 min	2h45	12€	+	++
Carpooling (as passenger)	40 min	2 h	35 min	3h15	14€	+	++
Air	1h40	3h15 (with connections)	1h25	5h20	200€	–	–

Adapted from UIC (2016)

Table 2.4 Transport mode that would have been used in the last trip if carpooling were not available

Mode	Driver (%)	Passenger (%)
Private vehicle	67	16
Train	14	42
High-speed train	10	27
Bus	1	2
Airplane	1	1
No trip	8	12

Adapted from ADEME/6T (2015)

UIC (2016) showed that direct competition exists between carpooling and bus services, with a competitive advantage for carpooling in terms of both pricing and duration of the trip. Carpooling offers an attractive substitute to railway transportation when passengers want to reduce costs and are ready to compromise with the duration of the trip.

Thirdly, migration from collective services to carpooling is also confirmed by surveys among carpooling users, which show that carpooling users would use collective transportation if carpooling were not available. According to the survey by ADEME/6T (2015), this is the case for passengers (72%), but also for drivers (26%).[8] ADEME/6T (2015) even distinguishes the type of collective transport mode that would be used, as summed up in Table 2.4.

Fourthly, quantitative analyses have been conducted on the impact of carpooling services in collective services in France. An estimate is provided by CGDD (2016), according to which the number of pkm traveled by train would have been 52.3 billion, compared to 49.2 billion if carpooling had not been available. That is a reduction of 3.1 billion pkm, which represents approximately 6% of the actual 49 billion train pkm in France in 2015.

ADEME/6T (2015) indicated that each vehicle-km traveled by carpooling reduced the use of trains by two pkm (ADEME/6T 2015: 74). If the total number of vehicle-km in carpooling is estimated to be 2.5 billion in 2015,[9] this means that the total number of pkm traveled by train would have been reduced by 5 billion, around 10% of the total pkm in 2015.

Managers of bus services in Spain claim a similar impact in their long-distance operations. Bus services are popular in Spain among low-income passengers, while rail services, particularly high-speed services, tend to be more popular among high-income passengers. CONFEBUS, the Spanish trade association of bus companies, claims that carpooling has caused a 20% reduction in the number of long-distance bus passengers (BlaBlaCar 2016).

[8]These figures are confirmed in the survey in Shaheen et al. (2016, p. 12), as 65% of the passengers would use collective transportation, while 88% of the drivers would use a personal car.

[9]This figure is the result of multiplying 8 million annual trips by 320, the number of average km in carpooling trips in France (as in CGDD 2016).

In conclusion, the existing evidence shows that a large volume of passengers are migrating from long-distance collective services (train and bus) to carpooling. This is the main customer base for carpooling services, with a reduced volume of drivers becoming passengers in someone else's car.

2.4.4 Efficient First-Mile–Last-Mile Mobility and Suitable Distances

A fundamental factor in the development of long-distance carpooling is the facility for passengers to arrive at the location where drivers pick them up, and then reach their final destination point from the location where drivers drop them off. Carpooling faces the same first-mile–last-mile challenges as traditional long-distance transportation modes.

It has been confirmed that drivers are not usually ready to make significant detours from their prearranged route to pick up or drop off passengers. According to the survey conducted by ADEME/6T (2015), 36% of drivers did not detour at all, 35% made a detour of less than 2 km, 38% a detour of between 2 and 5 km, and 27% a detour of more than 5 km. The low price of the service might explain this behavior (see Sect. 2.4.2).

Consequently, stable meeting points are emerging as "hubs." In France, the most common locations for pick-ups are the train stations (36% of cases) and car parks in specific points such as commercial areas (29%) (ADEME/6T 2015: 64).

Of utmost relevance, therefore, is the way passengers reach the hubs. It has been confirmed that passengers mostly rely on local collective transportation to reach the hub (44% of passengers), while 22% have been transported by car by a third person, 20% have reached the meeting point on foot, 6% have driven their own car to the meeting point, and only 2% have been picked-up by the driver in their own location (ADEME/6T 2015: 64).

In Europe, population tends to be concentrated in dense urban areas and local collective transport services are usually well developed. As a result, passengers can easily reach the emerging carpooling hubs either by collective transportation or even walking.

Trip distance is also a key geographic parameter for the use of carpooling, as road transportation tends to be less competitive than air or high-speed rail transportation trips above certain distances. The popularity of carpooling seems to diminish as distance increases in such a way that carpooling has a very small presence on very long-distance journeys (above 500 km) (ADEME/6T 2015: 61). The comfort of railway services compared to private vehicles (more space, possibility to walk around the train, onboard services, etc.) seems to be the explanation for this. This fact also explains why carpooling has a limited effect on air transportation, as flights usually cover greater distances, particularly those above 500 km (UIC 2016).

2.5 Regulatory Challenges

2.5.1 The Platform as an Intermediary

The first regulatory challenge around transport platforms is to define their legal status. In particular, the debate has focused on whether they provide an intermediation service using digital technology, or whether they really provide a full transportation service, which often requires a license and full liability for the provision of transport services to passengers.

Platforms in multilateral markets facilitate interactions between third parties. The specific business model of each platform may vary, but platforms connect service providers with users, allowing the internalization of a network externality. Platforms are not designed to provide their own accommodation or transport services, but to facilitate the contracting of services provided by third parties.

However, the intermediation service provided by transport platforms is particularly powerful. As described along these pages, platforms do not merely use technology to make the request for a transport easier—they create network effects by coordinating independent, isolated vehicles. In this way, a new, multi-sided market is created and great efficiencies are generated.

Scale is necessary to create and distribute the efficiencies around transport platforms. It is necessary to attract a large number of service providers (drivers in the case of BlaBlaCar) and passengers. Such a scale cannot be built by a platform forced to be the owner of the vehicles, the employer of the drivers and fully liable for the transport services. Growth would be too slow. On the contrary, growth can be faster if the platform can pool together the assets and work of independent service providers. This is the way BlaBlaCar has grown to create efficiency by internalizing indirect network effects around large pools of drivers and passengers.

BlaBlaCar has grown particularly quickly as it attracted non-professional drivers to the market. Drivers are individuals making their own long-distance trips and offering the empty seats in their vehicles. In the case of Uber or Lyft, drivers are individuals using their own vehicles, often for just a few hours a week to earn extra revenue (uberPOP), even if professional drivers with private hire vehicle (PHV) licenses are common in the USA and currently support most of Uber services in Europe.

Transport platforms have disrupted traditional transport companies that benefited from an absence of network effects (taxi) or relatively smaller network effects (railways and bus companies). Such companies have often complained about the regulatory burdens they face (licenses, taxes, labor conditions, etc.) while platforms (or more precisely the non-professional drivers using the platform) often had no license and paid not taxes, etc.

Traditional transport providers have challenged the legality of transport platforms before courts all around Europe. This has certainly been the case with Uber, but also with BlaBlaCar in a specific market: Spain. It has been argued that platforms are not mediating the provision of services by third parties, but are actually the service providers, and they are often providing services without the required licenses and without meeting other regulatory obligations, therefore competing under unfair terms.

Two legal regimes are in conflict. On the one hand, transport regulation tends to be local and restrictive in terms of licensing (often under exclusive rights), public service obligations, pricing, etc. On the other hand, digital services are global and ruled by a flexible regulation adopted to foster the growth of digital services. This is the case of the USA and the exemption of liability in section 230 of the Communications Decency Act; and also the case of Europe, where Directive 2000/31 (e-Commerce Directive) established freedom to provide services framework according to which digital services are ruled by the law of the country of establishment of the service providers (France, in the case of BlaBlaCar), with no previous authorization, and with a strong protection against restrictions introduced by the recipient State. In the case of transport platforms, which regime would apply?

This clash of legitimacies has been adjudicated by the Court of Justice of the European Union (CJEU) in a recent judgment involving Uber.[10] The CJEU established the two principles for the regulation of transport platforms.

Firstly, the CJEU has confirmed that transport platforms are intermediaries and not the providers of the underlying transport service. The Court has repeatedly differentiated between the non-collective urban transport service and the service offered by Uber. The CJEU confirms that "*passengers are transported by non-professional drivers using their own vehicle*" (para. 38), while the platform, in this case Uber, provides a different service, which the CJEU repeatedly qualifies as "intermediation service" (see, for example, para. 40 or the final ruling).

However, the CJEU is correct when it states that Uber "*is more than an intermediation service consisting of connecting […] a non-professional driver using his or her own vehicle with a person who wishes to make an urban journey*" (para. 37). Thus, the platform is not limited to the transfer of information between the driver and the passenger (para. 35).

The CJEU appropriately identifies that "*the provider of that intermediation service simultaneously offers urban transport services, which it renders accessible […] and whose general operation it organizes for the benefit of persons who wish to accept that offer in order to make an urban journey*" (para. 38). The platform allows each individual driver to benefit from coordination and the effects of operating as a network: shorter waiting time, fewer empty drives, consequent cost reduction, etc. This is precisely the added value offered by the platform. The CJEU succeeds in identifying that the mediation of platforms in multilateral markets transforms the market by facilitating that a previously fragmented demand functions as a structured network. In any case, this effect of the service provided by the platform does not mean that the platform is the provider of the transport service. The transport service provider remains to be the driver. The platform provides an

[10]Judgment of December 20, 2017, *Elite Taxi/Uber*, C-434/15, ECLI:EU:C:2017:981.

intermediation service. In this way, the CJEU has validated the model of industrial organization carried out by platforms in multilateral markets.

Secondly, and against the interest of transport platforms active in Europe, the CJEU has decided to exclude the application of the Directive on e-Commerce and in general the protection of the European Union legislation on freedom to provide services. Transport platforms are ruled by the transport legislation of each member state.

The CJEU understands that the intermediation service provided by Uber, being a differentiated service and not the transport service, is "*indissociably linked*" to it (para. 41). The key reason is that "without which those drivers would not be led to provide transport services" (para. 39). Furthermore, the Court states that the transport service is the "*main element*" (para. 40). As a result, the Treaty's exceptional regime on transport exercises a *vis attractiva* that excludes the application of the free service provision regime (Article 56 TFEU), the Services Directive, and even the e-Commerce Directive.

In conclusion, the CJEU confirms that transport platforms provide intermediation services and not the underlying transport service, thus validating a legal qualification that allows the development of this model of industrial organization in Europe. However, the CJEU has excluded the application of the European Union regime on freedom to provide services.

The two main principles in this judgment can be perfectly transposed to the intermediation service provided by BlaBlaCar. It is legally an intermediation service of a transport service provided by non-professional drivers, and despite the use of technology, it does not benefit from the regime defined in the e-Commerce Directive, so it is governed by the national legislation on transport.

For BlaBlaCar, the only legal challenge to its activity was decided in Spain by Commercial Court no. 2 in Madrid on February 3, 2017, after an unfair competition lawsuit from the local coach association. In Spain, long-distance coach services are operated under exclusive rights granted after a tender organized by the relevant authority (the State or the region), packaging routes so that profitable routes connecting cities cross-subsidize loss-making routes to small villages. Coaches in Spain have higher passenger volumes than in the rest of Western Europe, as railways networks are more limited. The Spanish coach association understood that BlaBlaCar was breaching the exclusive rights of its members, diminishing the number of passengers and threatening the financial viability of the companies and the provision of services to smaller towns and villages.

However, the Commercial Court in Madrid decided that BlaBlaCar was only mediating in the provision of the carpooling service, and furthermore, that the underlying carpooling services mediated by BlaBlaCar are private services that can be provided with no license as the price is below EUR 0.19/km, the legal reference to reimburse expenses to civil servants when traveling with their own car; that is, the service is being provided with no profit.

In the USA, non-profit carpooling services are specifically considered in some state regulations such as California, and the drivers do not require a license. Such services are defined as "*transportation of persons between home and work locations or of persons having a common work-related trip purpose in a vehicle having a seating capacity of 15 passengers or less, including the driver, which are used for the purpose of ridesharing, as defined in Section 522 of the Vehicle Code, when the ridesharing is incidental to another purpose of the driver. [...] This exemption does not apply if the primary purpose for the transportation of those persons is to make a profit. 'Profit,' as used in this subdivision, does not include the recovery of the actual costs incurred in owning and operating a vanpool vehicle*" (Public Utility Code §533(h)).

2.5.2 Level Playing Field

The debate on the legal nature of transport platforms is closely linked to the debate on whether a level playing field exists between traditional transport companies, transport platforms, and service providers using platforms. Sector-specific regulation imposes major restrictions on traditional players, such as price regulation and public service obligations on coach companies and railway undertakings (frequencies, services to low-density areas, etc.), as well as fixed prices on taxis. Tax, social security, and labor law also generate high costs in the operation of traditional transport services.

Non-professional service providers, on the contrary, are often exempted from sector-specific obligations (price regulation, public service obligation, etc.) and even of the most relevant general obligations on taxation, social security, and labor law. Non-professional providers often do not charge VAT with their fees, and they do not pay corporate taxes (or even personal income taxes), etc.

Transport platforms do pay taxes as a regular corporation. However, taxes are not always paid in the jurisdiction where the underlying transport service is provided. Taxes on the value appropriated by the transport platforms are not always reinvested in the jurisdiction where the transport service is provided.

As new transport services are substituting transport services provided by traditional operators, a relevant decrease in taxes and social contributions from the traditional operators is taking place, and such public revenue will not be always compensated by new revenue from new transport service providers. Failure to pay taxes reduces the amount of finance available for transport infrastructure and, at the same time, reinforces the competitiveness of non-professional service providers against traditional transport (Finger et al. 2017).

2.5.3 Relationship Platform/Service Providers

Experience shows that there are frequent tensions between transport platforms and service providers mediated by them. This has not been the case with BlaBlaCar. On the contrary, tensions have been more common around other transport platforms such as Uber and Deliveroo.

Such tensions are more common as drivers professionalize and work a significant number of hours to make use of a platform. Platforms tend to aggressively reduce maximum prices as to increase the attractiveness of the services they mediate and grow larger and larger pools of users. It is often the case that larger pools of users reduce empty drives and increase the load factor, reducing the cost of the provision of the service for the drivers. Lower prices can increase revenue for the drivers. In any case, the relationship between small service providers and large platforms is becoming increasingly unbalanced. The term "Uberization" is being used to describe the weak position of workers in the digital age.

Making the drivers employees of the platforms is not the best solution to protect these individuals while making available for the whole society the efficiencies derived from network effects in multi-sided markets.

An alternative that is—we understand—more appropriate to channel the growth of this model of industrial organization is to introduce specific legislation to strengthen the guarantees for service providers, always within the framework of a commercial relationship for the provision of services, and specifically of a legal nature of an intermediation or brokerage contract. We understand that, in this framework, instruments to protect service providers could be introduced. Some of them could be inspired by some protection instruments of the labor relationship (insurance against accidents, vacations, minimum payments, payments for termination of the contract, etc.), always under strict principles of necessity, adequacy, and proportionality.

2.5.4 Relationship Platform/Final Users

The relationship between the platforms and the receivers of the transport services can also raise problems. In order to face these challenges, we propose to rigorously apply the consumer protection legislation in provision of the intermediation service.

In this framework, the complexity of the business model of platforms in multilateral markets seems to recommend extreme transparency in the identification of the different actors, their obligations and liabilities. Furthermore, we propose that the platforms assume a certain responsibility in the provision of the main service, even if it is subsidiary to the responsibility of the service provider.

It is also proposed to impose more control on the platforms in the filtering of access, especially of suppliers (identity verification, provision of qualifications, insurance, etc.). The main platforms have voluntarily developed measures along these lines to increase the security of transactions and thus increase volumes. However, this voluntary intervention has a limit because the more responsibility the platform assumes, the greater the risk that the public authorities will consider that their activity is not limited to mediation, but that they become owners of the provision of the main service, thus being obliged to assume full responsibility for it. This is a paradoxical situation, which in no way benefits the users.

2.5.5 Market Power

In relation to the consolidation of a "winner-take-all" model, it is convenient to identify the reality in the market before intervening in markets that are in their infancy and in which there is great uncertainty about their evolution.

The application of antitrust rules to platforms, either in the context of a concentration or in the framework of a procedure for abuse of dominant position, requires a careful approximation.

In the first place, the definition of the relevant market must be correct. Multilateral markets can be analyzed either as a whole or as a sum of different markets. It has been pointed out that it seems more appropriate to analyze transactional multilateral markets as a single relevant market, as opposed to the possibility of independently analyzing each side of the market in non-transactional ones. However, there is a risk that the analysis may be excessively narrowed if digital markets are considered in isolation from competition in traditional physical markets. For example, it is not possible to analyze the market for food delivery platforms without taking into account the traditional distribution market. In any case, it is important to ensure that the analysis takes into account the existing connection between the different markets linked by the platform.

By analyzing the most popular platforms, we can distill some lessons. In accommodation, a large number of platforms are competing with the leader, Airbnb; for example, HomeAway has a significant presence in the USA. In urban transport, there are alternatives to Uber in many jurisdictions. In China, Didi has a greater market share than Uber. Even in the USA, Lyft has a significant and growing presence (24% market share in San Francisco). In long-distance transport, BlaBlaCar has competitors like Amovens, although it is true that the former accounts for more than 90% of trips in the most mature markets in Western Europe.

It can be advanced that, in thick markets (those with a high number of transactions), with few barriers for the parallel use of several platforms (multi-homing), and in which network externalities get exhausted after a certain threshold that can be replicated (e.g., waiting times for the arrival of the vehicle under one minute) there would be no tendency to "winner-take-all." This could be the case in the accommodation and urban transport sectors. In interurban transport, especially on lower-density routes, there does seem to be a tendency to "winner-take-all" as there are a fewer transactions.

There is a growing tendency to question whether the application of competition rules is sufficient to tackle the concentration of power in the main transport platforms, and in general digital platforms.

Identifying the market power in the network effect, a mandatory interoperability between platforms is considered. This has been the way in which the market power in the traditional networked industries, such as telecommunications, has been addressed: mandatory network interconnection. The positive network externality was extended to all operators. This approach seems possible (albeit technically complicated) for operating systems (apps, etc.).

Those people who have identified the market power in the availability of large volumes of data (big data) and the efficiencies derived from their use have proposed a public intervention aimed at sharing data, or at least that users can carry their data from one platform to another. This intervention builds on the regulatory experience of telecommunications and the figure of number portability and seems especially appropriate for social networks, but it does not seem to be described as having a determining effect in relation to transport platforms.

In any case, it does seem clear that public intervention in relation to platforms in general, and transport platforms in particular, is set to increase in the coming years, either to reduce concentration and increase competition, or to discipline the market power of the main platforms.

2.6 Conclusions

Transport platforms such as BlaBlaCar, Uber, Lyft, Didi, and Deliveroo are leading examples of a new industrial organization model: platforms in multi-sided markets. Online platforms use technology to facilitate new interactions between transport providers and passengers. As large pools of service providers and passengers use the platform, indirect network effects are generated, internalized by the platforms, and distributed across the ecosystem. The larger the pools of users, the more relevant the efficiencies, igniting a virtuous cycle that reinforces the business model.

BlaBlaCar is a leading example of the potential of transport platforms. BlaBlaCar has revolutionized long-distance transport in Europe offering attractive carpooling services. Non-professional drivers share the empty seats in their vehicles when traveling, for a price that covers only variable costs. Low prices attract passengers. As the platform grows, passengers benefit of more routes and more frequencies. Drivers benefit from more passengers and higher revenue. However, low prices and high frequencies detract passengers from public transportation.

From a regulatory perspective, it is important to recognize that the efficiencies created by transport platforms are only possible if large numbers of users—both drivers and passengers—pool together in the platform. Such large volumes cannot be the result of platforms growing their own fleets of vehicles, driven by their employees. Growth, indirect network effects, and efficiencies can only be obtained by pooling together idle resources owned by third parties, often non-professional drivers, operated by the owners of such resources. Platforms can only grow if it is recognized that they merely intermediate and that they do not provide the underlying transport service, they do not own the vehicles, drivers are not employees, and the platform is not fully liable for the service provision.

In any case, platforms such as BlaBlaCar play a leading role as intermediaries. They do not merely use smartphones to connect drivers and passengers. They coordinate supply as to make it work as a network. They manage the tools to create a safe contracting environment (evaluations, payment systems, insurance, etc.).

Finally, they determine the principles for the distribution of internalized network externalities across the ecosystem (drivers, passengers, and the platform itself) in the form of prices and commissions.

Regulation must respect the DNA of digital platforms in multi-sided markets, that is, their role as intermediaries. A level playing field must be defined for the competition between traditional transport companies and shared transport. Taxes, social contributions, and other regulatory obligations have to be distributed evenly. Adequate protection must be provided, both for transport service providers and the passengers using the platform. And finally, market power in potentially winner-take-all markets has to be supervised in order to meet the general interest objectives in an industry as relevant for the general interest as transportation.

References

ADEME/6T Bureau de recherche (2015) Enquête auprès des utilisateurs du covoiturage longue distance. ADEME

ATEMA/ADEME (2010) Caracterisation de services et usages de covoiturage en France: quels impacts sur l'environnement, quelles perspectives d'amelioration?

Beck J (2016) The decline of the driver's license. The Atlantic, 22 Jan. http://www.theatlantic.com/technology/archive/2016/01/the-decline-of-the-driverslicense/425169/. Accessed 30 Dec 2016

BlaBlaCar (2016) Sobre el juicio a BlaBlaCar. http://www.blablacar.es/blablalife/novedades/juicio-blablacar. Accessed 30 Dec 2016

BlaBlaCar (2017) BlaBlaCar—Le Site Du Covoiturage En Europe. http://www.covoiturage.fr. Accessed 8 Feb 2017

Botsman R, Rogers R (2010) What's mine is yours: how collaborative consumption is changing the way we live. HarperCollins, New York

CGDD (2016) Covoiturage lingue distance: état des lieux et potentiel de croissance. Etudes & Documents du Commissariat Général au Développement Durable, no. 146, May

Chan ND, Shaheen SA (2012) Ridesharing in North America: past, present, and future. Transp Rev 32(1):93–112

Damme E, Filistrucchi L, Geradin D, Keunen S, Klein T, Michielsen T, Wileur J (2010) Mergers in two-sided markets. NMA reports. Netherlands Competition Authority: 1–183

European Commission (2016) Fifth report on monitoring development of the rail market. COM (2016) 780

Evans DS (2011) Platform economics: essays on multi-sided businesses. CPI, New York

Evans DS, Schmalensee R (2016) Matchmakers. The new economics of multisided platforms. Harvard Business Review Press, Boston

Finger M, Bert N, Kupfer D, Montero J, Wolek M (2017) Infrastructure funding challenges in the sharing economy. European Parliament, Policy Department for Structural and Cohesion Policies, Brussels

GART/UTP (2014) Covoiturage et transports collectives: concurrence ou complementarite sur les deplacements longue distance? In: Congrès GART-UTP, July 2014

Godillon S (2016) Complémentarité et/ou substitution entre le covoiturage et les transports collectifs? Les Rencontres de la mobilité intelligente, Paris, 26–27 Jan 2016

Kelly K (2007) Casual carpooling—enhanced. J Public Transp 10(4):119–130

Libert B, Beck M, Wind J (2016) The network imperative. How to survive and grow in the age of digital business models. Harvard Business Review, Boston

Mazzela F, Sundarajan A (2016) Entering the trust age. https://www.blablacar.com/trust. Accessed 30 Dec 2016

McAfee A, Brynjolfsson E (2017) Machine, platform crowd: harnessing our digital future. W.W. Norton & Company, New York

Minett P, Pearce J (2011) Estimating the energy consumption impact of casual carpooling. Energies 4(1):126–139

Montero J, Finger M (2018) Platformed! Network industries and the new digital paradigm. Compet Regul Netw Ind 18(3–4)

Munger M (2015) Coase and the 'sharing economy'. In: Veljanovski C (ed) Forever contemporary: the economics of Ronald Coase. Institute of Economic Affairs, London, pp 187–208

OECD (2011) Two-sided markets. Policy Roundtables. https://www.oecd.org/daf/competition/44445730.pdf?TSPD_101_R0=73b5d4173d7cef1de62653739e10c83eice00000000000000029fcb380ffff00000000000000000000000000005a942a0d0055dd7004. Accessed 30 Dec 2016

Oudghiri R, Brunet L (2013) Les Français et les pratiques collaboratives. Qui fait quoi? Et pourquoi? IPSOS Public Affairs et ADEME, 122 p

Parker G, van Alstyne M, Choudary S (2016) Platform revolution. W.W. Norton & Company, New York

Pratsch L (1975) Carpool & buspool matching guide, 4th edn. US Department of Transportation, FHWA, Washington, DC

Rochet JC, Tirole J (2003) Platform competition in two-sided markets. J Eur Econ Assoc 1(4): 990–1029

Rochet JC, Tirole J (2006) Two-sided markets. A progress report. Rand J Econ 37(3):645–667

Shaheen S, Stocker A, Mundler M (2016) Online and app-based carpooling in France: analyzing users and practices—a study of BlaBlaCar. Transportation Sustainability Research Center, University of California, Berkeley

Steer Davies Gleave (2016) Comprehensive study on passenger transport by coach in Europe. Report prepared for DG MOVE. European Commission Directorate-General for Mobility and Transport, Brussels

Sundararajan A (2016) The sharing economy: the end of employment and the rise of crowd-based capitalism. MIT Press, Cambridge, MA

TCRP (2012) Ridesharing as a complement to transit. http://www.tcrponline.org/PDFDocuments/tsyn98.pdf. Accessed 30 Dec 2016

UIC (2016) Status of the competition for long distance trips at the aggregated level. Carpooling, bus, train, private car and air competition on long distance trips

Weiner E (1999) Urban transportation planning in the United States: an historical overview. Praeger Publishers, Westport, CT

Juan Montero is a professor in the Administrative Law Department of the UNED University in Madrid, where he lectures on economic regulation, and a part-time professor at the Florence School of Regulation of European University Institute in Florence. For the last 20 years, his research area has been the regulation of the traditional network industries, in particular transport, telecommunications, and postal services (independent regulators, access regulation, public service obligations, etc.). In recent years, his research has expanded to the new network industries (online platforms in the sharing economy, social networks, etc.), as the indirect network effects created by platforms raise similar regulatory challenges as the direct network effects created by infrastructures.

Chapter 3
Regulation of TNCs in Latin America: The Case of Uber Regulation in Mexico City and Bogota

Maria Lorena Puche

Abstract Over the past several years, the development of new information and communication technologies (ICTs), and the widespread increase of smartphone ownership, has enabled the creation of new transportation modes. Within these solutions, e-hailing services, facilitated by Transportation Network Companies (TNCs), have emerged to disrupt the taxi industry worldwide and pose major regulatory challenges for authorities and policymakers. Many specialists have addressed e-hailing regulation in cities across Europe and the USA. However, little research has been done focusing on Latin America. Therefore, this paper's objective is to explore how authorities are regulating e-hailing services in Latin American cities. Moreover, should these regulations be based on maintaining fair competition regarding traditional taxis? We attempt to answer these questions using of a conceptual framework based on taxi regulation, as this seems to be how authorities are considering regulating e-hailing services. We then develop a comparative case study of Uber regulations in Mexico City and Bogota. A discussion and analysis follow, regarding the Uber regulations approved in both cities. Finally, we conclude the paper with a summary of the findings, and suggestions for future studies.

Keywords Regulation · TNCs · E-hailing · Uber · Latin America

3.1 Introduction

Recent advances and widespread use of information and communication technologies (ICTs) have supported the creation of new mobility modes. These new solutions have exploited a breach created by low-quality public transportation services, including taxis. Additionally, young people's tendency to not drive nor own cars, together with their increased use of smartphones and the Internet, has also contributed to the increase in new modes of mobility (Cannon and Summers 2014).

M. L. Puche (✉)
Independent Researcher, Miami, USA
e-mail: marialorena.puche@gmail.com

M. Finger and M. Audouin (eds.), *The Governance of Smart Transportation Systems*, The Urban Book Series, https://doi.org/10.1007/978-3-319-96526-0_3

In particular, satellite-based dispatch systems for ride services (also known as electronic hailing systems, e-hailing, or ride-sourcing), provided by transportation network companies (TNCs) such as Uber and Lyft, have taken advantage of this by using smartphone applications and Internet connections (Yussoff et al. 2015). Uber and Lyft are revolutionizing the taxi market by allowing drivers to more easily find passengers, spend less time in transit between rides, and potentially maximize the number of passengers they carry per day (Geloso and Guénette 2014). They have also sparked the interest of customers for offering a better service at lower costs and with greater reliability. Silverstein (2014), compared Uber and taxi rates in 21 large US cities and concluded that Uber was cheaper than a taxi everywhere, except in New York and Philadelphia. Rayle et al. (2014) found that introducing Uber and Lyft apps in the San Francisco market considerably reduced waiting times and showed that customers prefer e-hailing services because of simplified payment and the ease of calling a car. Finally, Uber users can rate their drivers, encouraging superior service to boost reputational scores (Geradin 2015).

However, the rapid expansion of e-hailing in the taxi industry began to outpace the current regulations in countries where it operates. TNCs have faced criticism, as well as fundamental legal threats, for unfairly competing with taxi drivers by entering the market without following existing regulations. In major cities, such as New York, capping the number of taxis allowed on the road is a common regulatory tool. This practice has driven the price of medallions as high as US$1.3 million per unit. However, the average price of New York City taxi medallions has suffered a 17% decline since a peak in 2013, due to Uber's entry into the market. In Boston and Chicago, medallion prices also have declined by 17 and 20%, respectively (Barro 2014). Uber has also faced criticism in terms of safety, such as the protection of private data and the lack of appropriate insurance (Hanks and Alexander 2014). In some places, suspension or outright bans are the first reaction of public authorities to the appearance of e-hailing services (Grand and Khosla 2015). In countries such as Spain, Thailand, Germany, Vietnam, Amsterdam, and the Netherlands, Uber has been banned for operating without the necessary licenses (Wall Street Journal 2014). On the other hand, TNCs usually respond by simply continuing operations, despite any ban or suspension issued by authorities. For example, Uber continued its operations in Paris, despite judicial and police interventions. Uber only suspended service in the city after the arrest of two Uber executives (Edelman 2015).

Consequently, many scholars have called for developing new regulatory frameworks (Cannon and Summers 2014; Harding et al. 2016). Edelman and Geradin (2015) support the idea of creating a level playing field. They think that policymakers should embrace the efficiency that technical platforms such as Uber provide and remove those requirements and protectionist rules that benefit taxi license owners more than customers. Moon (2015) presented e-hailing as an opportunity to retrofit taxi regulations into current markets. Others, such as Darbéra (2015), suggested developing regulations to avoid a monopoly in this new market. Strong (2015) suggested regulations as a means to reach environmental policy aims, while others suggested using regulation to mitigate social problems, such as employment and discrimination in access to transportation (Rogers 2015). Other authors, such as Farren et al. (2016), investigated the benefits and costs of completely deregulating the taxi industry.

Most of the studies mentioned above have been done in cities across Europe and the USA. Little research has been done in Latin American cities concerning how Uber is regulated, as well as the future of e-hailing services in this region. With high urbanization rates that average 80%, lack of investment in transportation infrastructures, low-quality of taxi services, high levels of congestion, and growing use of the Internet at a rapid rate, the region appears to be the perfect ground for e-hailing services to develop (Audouin and Neves 2017; Harrington 2012). Indeed, Uber has been rapidly and aggressively expanding since 2013 in Latin America, totaling 45 million trips just during August 2016 (Newcomer 2016). Therefore, the e-hailing Latin American market must not be ignored (Audouin and Neves 2017), as it is currently Uber's fastest-growing region (Uber Estimate 2018).

This paper will explore how governments can regulate e-hailing services to maintain fair competition with traditional taxis. We attempt to fill this research gap by exploring how some major Latin American cities have dealt with the rise of e-hailing services to, identify the main challenges that policymakers should address to regulate e-hailing services in the region. For that objective, we adopted a comparative case study design, as it is best suited for when it is difficult to separate the phenomenon of interest from the context (Yin 2009). We chose to look at Bogota and Mexico City because both are major cities in Latin America. They were the first cities in the region where Uber launched its service in 2013. Additionally, Mexico City became the first Latin American city to regulate Uber in mid-2015. Uber became regulated in Bogota at the end of 2015. The result in each case was completely different. Unlike in Mexico City, Uber did not accept the regulatory framework proposed in Colombia and is still facing different forms of resistance from public authorities and legal threats. The cases were produced with documentation data and semi-structured interviews conducted with key stakeholders involved in the project. This approach is in line with case study research strategy, as it allows the researcher to use multiple sources of data (Yin 2009). The rest of the paper is constructed as follows: In Sect. 3.2, we present a conceptual framework for analyzing Uber regulation. In Sect. 3.3, we present the case studies. In Sect. 3.4, we analyze the cases, using our conceptual framework. We present a conclusion in Sect. 3.5 and ultimately propose some leads for further research linked to e-hailing regulation.

3.2 Conceptual Framework

The objective in this section is to propose a conceptual framework to analyze how Uber and similar companies are regulated. We begin with a look at pertinent literature about the taxi regulatory framework that has governed the industry for decades. Although Uber is seen as a digital platform for connecting people (Badger

2014), rather than as a taxi company, it provides a solution for people's transportation needs that, in the view of the individual transportation sector, is quite similar to the service offered by taxis. A recent decision by the European Court of Justice (ECJ) confirmed the need to regulate Uber as a taxi, and not as a technology company. This decision was made mostly on the basis that the services provided by companies like Uber are "*inherently linked to a transport service*" and must be classified in the field of transportation (Bowcott 2017). Therefore, we present a conceptual framework building on taxi regulation, as it seems to be how policy-makers are considering regulating Uber.

Taxis are intended to provide vehicles and driving services for passengers (Li 2016). They play an essential role within the context of urban transportation systems, as a necessary component for supplying a public transportation function in accordance with public demand (Aarhaug and Skollerud 2014). There are three major market segments: the taxi rank, in which places are designated for taxis to wait for passengers, and vice versa; the hailing segment, in which a cruising taxi on the street can be hailed; and the pre-booked segment, in which consumers call a dispatching center (Salanova et al. 2011). According to OECD (2016), taxi market shares are linked to household incomes, costs of car ownership, and the availability of alternatives modes of transportation.

There are several arguments used for and against taxi regulation (Aarhaug and Skollerud 2014). To the growing ranks of free-market economists in 1970, taxi regulations were an example of convoluted state regulation that stifled competition and innovation (Harding et al. 2016). On the other hand, authors such as Buckley (2015) stated that taxi regulations are necessary to correct market failures that have occurred over time. There are three taxi market failures: asymmetrical information; destructive competition, leading to low quality of service; and externalities, such as pollution and congestion (Cohen and Sundararajan 2015).

In most cities, taxis are regulated following a standard regulatory framework that has been in place since the early 1930s (Harding et al. 2016). Most regulations, with some degree of variation, usually include the following three elements: monopoly rights, entry conditions, and fare controls (Beesley 1973). Regarding the first element, monopoly rights, authorities granted exclusivity rights to the taxi companies. Beesley (1973) stated that such rights could be given through two mechanisms. The first mechanism would be an exclusive franchise to organize taxi services in a geographically limited area. According to OECD (2016), a franchise is granted for a set time period in most cases. When the time expires, they are up for rebidding. Franchise systems prevent other taxi operators from picking up passengers in the franchise zone but allow drop-offs, which lead to unbalanced trips for those entering the zone from outside. The second exclusivity mechanism grants rights to a particular mode of operation (rank, hail, or dispatching center). However, as Schaller (2007) shows, it is difficult for regulators to measure productivity and thereby grant exclusivity rights, since the usage of public space (such as bus lane) has not been measured as a competitive advantage. The conclusions are that the dominant industry will obtain the maximum benefit with a small fleet and high prices (Salanova et al. 2011).

The second element of entry conditions includes quality and quantity controls. Quality regulation consists of vehicle control (age, type, and appearance), as well as driver and operator standards, such as fit-and-proper-person tests, uniforms, and route knowledge. Although quality regulation seeks to ensure passenger safety and minimum service standards, they can also have anti-competitive effects if they are set at unduly high levels (OECD 2007). Quantity control refers to limiting the number of taxis in a specific jurisdiction, which is modified by analyzing latent and future demand, or local politics and vested interests (Harris 2002). Over time, entry regulations were managed by freezing the number of taxi licenses, without justifying why the current number of taxis was optimal. Most cities maintained the number of taxis at 1980 levels, while a few increased the number of licenses, according to GDP variations or other economic indicators. Both methods created an inefficient taxi market, with either more taxis, or fewer vehicles than needed (Salanova et al. 2011). Some arguments for limiting the entry into the taxi market include: preventing crowding at stands or in city centers; keeping profitability; protecting workers from longer trips and lower wages; and preventing overcharging (Nelson/Nygaard 2008). On the other hand, the main argument against entry conditions is that they create economic rent (Aarhaug and Skollerud 2014).

Following Beesley's (1973) classification, fare control is the last aspect of taxi regulation. The first type of fare control is based on controlling the overall level of fares charged, and the second type is based on setting a pricing structure that is often based on distance and time. Cairns and Liston-Heyes (1996) proposed a model of taxi service demonstrating that price regulation is necessary to have equilibrium between supply and demand. OECD (2016) noticed that fare regulation prevented gouging in the street hail market, due to the asymmetric information, notably concerning the available supply and the uncertainty potential competitor's fares. In a monopolist market without fare regulation, higher fares will satisfy lower demand with a smaller fleet, maximizing the benefit of the operator (Douglas 1972). On the other hand, the same market with regulated fares will operate with the same size fleet in that the marginal benefit is equal to the marginal cost (De Vany 1975).

Finally, Audouin and Neves (2017) used a similar approach to construct a framework building on taxi regulation focusing on quality, quantity, price regulation, and market-conduct regulation. Therefore, the approach adopted in this paper is not totally new and builds on something that has already been done. In the next section, we will review our two case studies.

3.3 Case Studies

The following case studies are intended to illustrate the first regulatory reactions of Latin American cities to the entry of Uber into the taxi market. We will first present the case of Mexico City; and then present the case of Bogota. Both cases will describe the existing taxi regulations, examine the regulatory landscape Uber faced, and look at the new regulatory framework adopted for e-hailing services.

3.3.1 Mexico City

Mexico City, the capital and the largest city in Mexico, had 21 million inhabitants in 2016 and is expected to have more than 23 million inhabitants by 2030 (United Nations 2014). The unmeasured growth in population during the twentieth century has resulted in an unprecedented urban sprawl, following an extensive car ridership pattern. The average time spent commuting between home and work is 40 min by subway or train; 50 min by bus; 27 min by taxi, and 35 min by car (CAF 2011). Urban sprawl, along with the lack of development of an efficient transportation network, has caused the rapid expansion of taxi services as the third most popular transportation option (Parametria 2013), with a growing vehicle fleet of 106,000 regular units and approximately 22,000 illegal units (Lopez 2012). Currently, the combination of a steadily growing economy, record-low inflation, and declining mobile prices are contributing to accelerating migration to mobile broadband services and increase in smartphone usage. By 2020, Internet subscribers in Mexico will have grown to 84% of the population (GSMA 2016) and the number of smartphone users is forecast to reach more than 67.49 million in 2020 (Statista 2018).

Taking advantage of this context, Uber started operating for first time in Mexico City in June 2013, resulting in unprecedented competition with the taxi industry and a challenge to the existing regulatory framework governing the taxi industry. In this metropolis, the federal government (state level) is in charge of granting concessions for providing transportation services. It also regulates fare policy decisions, public service supervision, route design, and other activities (Islas et al. 2011). In July 2014, the Mobility Law of 2002 was replaced by a new legal framework to regulate city transportation services. Since then, all drivers must have liability insurance. Additionally, drivers must approve training courses and assessments' processes established by the Secretary of Mobility (SEMOVI). Taxi regulations are based on exclusivity rights, quality, and price regulation. Regarding, exclusivity rights, providers offer a service under a concession awarded by the federal government and must meet the requirements set forth by transport authorities regarding operation territories, bases, concessions, and fees. Quality regulations include characteristics and requirements that the vehicle must satisfy (age, color, and identification), as well as those that drivers must meet, such as courses about driving, self-defense, and civility. Fares are also defined by the federal government, and all taxis must use a taximeter. It is important to highlight that the city government has not approved new taxi licenses, resulting in an uncertain market in which you can buy a concession for MEX$40,000–MEX$60,000 (USD$2150–USD$3200).

Once Uber entered into the Mexico City taxi market, as in many other cities worldwide, the taxi sector protested and lobbied the government to expel Uber. It was not until after two major demonstrations, in October 2014 and May 2015, in which the city's main roads and access to the International Airport were blocked, that the city government made this conflict a real priority in its agenda. On June 17, 2015, authorities organized working meetings, called Debate Digital CDMX, to discuss and regulate e-hailing services (García 2016). The primary purpose of these meetings was to create proposals that would allow for innovation in services offered by taxis but under a scheme of fair competition. These meetings were set up with the participation of conventional taxi operators; representatives from Uber, EasyTaxi, and Cabify; experts in mobility and technology; and NGOs. All stakeholders gave recommendations aimed at improving the conditions under which taxi companies operate, allowing for the coexistence of different types of taxi services in México City. They defined the aspects that should be considered in the agreement, as well as improvement conditions for traditional taxis.

Simultaneously, a federal and a local government agency gave their opinions in favor of regulating Uber and similar companies. Based on the principle of economic competition, and the freedom of citizens to choose their way of moving around the city, they made the announcement in favor of Uber regulation on June 4 and 15, 2014, respectively (García 2016). In July 2015, the city government announced an administrative agreement to allow Uber and similar companies to legally operate in the capital. This special administrative agreement was published in the Mexico City District Gazette No. 133 Bis. July 15, 2015. Therefore, Mexico City became the first Latin American city to regulate e-hailing services. There were several aspects included with these regulations. App platforms must be registered with the Secretary of Mobility; and drivers must pay MEX$1599 (almost USD$100) for an annual permit on each vehicle used for this service. Furthermore, vehicles must have an original value of no less than MEX$200,000 national currency (nearly USD $11,000) and fulfill some requirements (four doors, air conditioning, air bags, and seat belts). Finally, e-hailing companies must pay 1.5% of each ride to the Taxi, Mobility and Pedestrian Fund, created by the city government for public work projects related to mobility. There are also a few restrictions, such as the prohibition on receiving cash or prepaid cards as a payment. Drivers are also not allowed to sublease their vehicles, nor can they can have a set base or fixed site.

Uber has agreed with the proposed regulation and said that the new regulation "*makes Mexico a pioneer in recognizing in law that supply should respond directly to demand and the free choice of consumers.*" However, two years after this regulation was proposed, progress in its implementation has been quite slow and seems to be stuck. To date, the city has not completed registering the taxi apps' fleet, which is needed to launch the fund. Furthermore, there is no time schedule for the fund's creation (Torres 2016). Overall, TNCs are still operating with minimal conflicts against conventional taxis.

3.3.2 Bogota

Bogota is the capital of Colombia and is expected to have more than 11.6 million inhabitants by 2030 (United Nations 2014). Urban growth is characterized by a growth in housing in surrounding areas (urban sprawl) that has not been adequately followed by a concurrent increase in urban transportation infrastructure. The average time spent commuting between home and work is 40.5 min by car; 34 min by taxi; and 73 min by bus (CAF 2011). Commuting by bus takes almost twice as long as by taxi. Therefore, taxis play a significant role, not only within Bogotá's transportation integrated system as feeders, but also, in providing advantages to the riders regarding accessibility, time efficiency, and comfort (Ibañéz 2012). By 2020, Internet penetration in Colombia is expected to be more than 52.9% of the population (eMarketer 2016). Mobile devices are at the core of Vive Digital Colombia, a program overseen by the Ministry of Technology to achieve its four-year goal of "*widespread adoption of internet and the development of a nationwide digital ecosystem*" (eMarketer 2015).

In Colombia, the Ministry of Transportation (nationwide level) heads the transportation sector, including the regulation of taxis (CAF 2011) in terms of defining policies, operations, and fares, among others things. Taxi regulation is quite limited and based on exclusivity rights, quantity, quality, and fare regulations. The local government establishes the rules and regulations governing mobility in the city and taxi vehicle circulation. Regarding exclusivity rights, the ministry regulates taxis by granting indefinite authorization to taxi companies, as long as they fulfill all the required conditions and criteria to provide a good service. Quality regulations cover a number of specific requirements for taxi companies (insurance and no criminal records), drivers (criminal background checks, a minimum amount of driving experience, and a behavior course), and vehicles (age limits, taximeter, and vehicle identification). Quantity controls are enforced by controlling the number of vehicles. Adding vehicles to the taxi fleet can be done incrementally or by replacement. The first scenario occurs when the taxi company experiences growth in the number of vehicles, while the second occurs when companies replace one of its already registered cars. Local authorities cannot authorize new cars in taxis services unless the need for the service is determined as a result of a technical study described in the Law-Decree 172 of 2001 (Ibañéz 2012). To link a vehicle to the fleet of a taxi company, a contract must be formalized between the vehicle owner and the company, which is made official to the Ministry of Transport once the local transport authority issues an operation card, which authorizes the vehicle to provide taxi service. The national government establishes the general principles for fare calculations to control fares.

Since the early 1990s, Bogota has only allowed the entry of new vehicles into the taxi fleet by replacement, creating the right by replacement, also known as *cupo* (quota) (Decree Number 613-1993). This practice has created two issues. First, a black market has emerged to buy the *cupo*, the price of which depends on the market supply demand, because it can change without any type of regulation. Prices can fluctuate from COL$81 million to COL$95 million (about US$28,000–US $33,000). Second, the number of illegal taxis has been growing steadily throughout the years, due to the cost of the *cupo*. Additionally, the mayor of Bogota is responsible for calculating the rate of the taxi fare and other fees related to taxi operation (Decree No. 400, dated September 26, 2014).

Similar to Mexico City, Uber came into the Bogota taxi market in 2013, having conflicts with both taxi drivers and regulators. According to traditional taxi drivers, Uber is an unfair competitor. They state that Uber drivers and vehicles do not comply with regulations established by authorities. Traditional taxi drivers also argue that Uber drivers do not pay taxes, receiving greater profit from a much lower investment (Sanchez et al. 2016).

After many disputes between taxi drivers and Uber drivers, as well as taxi drivers' threats of strikes and blockades, Colombian President, Juan Manuel Santos, delegated to the Ministry of Transportation, the creation of a decree to regulate the service provided by Uber and similar companies. In November 2015, Decree 1079 was created, which unified all regulations for the transportation sector. In this decree, the services provided by Uber and similar companies were regulated as a new mode of individual public transportation categorized as luxury taxis. Consequently, there are two kinds of taxi services in Colombia: basic taxi service, which is the same as the existing traditional taxi service; and luxury taxi service, which can only be booked through an app or digital platform. Vehicles must be a black four-door truck or sedan, and comply with the following specifications: a line on the side; have GPS; have a passenger cabin with capacity to accommodate a minimum of five people and space for baggage with capacity no less than 0.40 m^3. Moreover, luxury taxis must be new or not more than seven years old. Finally, drivers cannot receive cash as a payment, and a minimum base fare must be established, that must be equal or lower than the one established for the basic taxi service.

However, Uber decided not to accept the regulations established by the national government and not to register itself as a luxury taxi company. Despite the Ministry of Transportation's many pronouncements regarding the illegality of Uber, and the fines that the local government imposed, Uber keeps operating and offering its services in Bogota. Although the number of citizens demanding Uber keeps increasing, conflicts between Uber and the traditional taxis keep arising. For instance, the traditional taxi sector has required the Ministry of Technology to block the Uber app in the country. However, the Minister has said that from a technological viewpoint, the app is perfectly legal.

The Bogota government, through the Secretary of Mobility, has been working to create new taxi companies that claim to offer the same quality of service as Uber but comply with the requirements established in the decree. In a recent interview, the Secretary of Mobility Juan Bocarejo announced that these new taxis companies would be operating in the upcoming months. Although they will work through a digital platform like Uber, the main difference is the fare system. Fares will be calculated using a base price established by the local government and depends on the time of day (higher during the rush hours) (Opina Bogota 2017).

3.4 Analysis

In the conceptual framework, we established the three elements that have been frequently used to regulate the taxi industry: exclusivity rights, entry conditions, and fare control. In both Mexico City and Bogotá, taxis have been governed through schemes based on these three dimensions. In both cases, the scope of the regulation is limited and has sometimes been interpreted in different ways. Moreover, despite the fact that (as in most regulated industries) consumers can complain to the service regulator, this does not apply in the case of the taxi industry. In both cases, there are no mechanisms established by regulators by which taxi users can report their dissatisfaction in an efficient way. Therefore, taxis are not interested in providing a better service. Since there are few public transportation options, taxi users are forced to accept a poor-quality service with high fares. As a result, the industry has been dominated by the taxi guild and authorities who have lost control and supervision over the operation of this sector. This has led to a lack of accurate data concerning the reality of taxis operating in urban areas.

With the emergence of Uber in Mexico City and Bogota in 2013, the traditional taxi industry started to face increasing competition from this new type of transportation service. Uber threatened to eliminate the traditional taxi industry′s stronghold. Since then, traditional taxis are facing more demands from their users in terms of quality, efficiency, comfort, and affordability. In both cities, the traditional taxi sector fought the emergence of Uber by lobbying at city government headquarters. At the same time, many users have shifted to Uber service due to lower fares, cleaner cars, and higher-quality service. Moreover, they lobbied in support of the company, playing an important role in pressuring local authorities to enact new regulations to govern Uber and other similar companies.

In Mexico City, Uber and similar companies were regulated as TNCs, under different requirements than those for traditional taxis. The main aspects of e-hailing regulation can be summarized as follows: App platforms and car operators must be registered with the Secretary of Mobility; the type of service should only be as TNCs; restrictions on the vehicle type (cost and environmental requirements); restrictions on payment method; and contribution of each ride to a mobility fund. Our analysis of the regulatory approach for e-hailing services in Mexico City found it to be based on exclusivity rights and entry conditions. As we saw, one way of granting exclusivity rights is based on a particular operation mode. In the regulatory framework of Mexico City, e-hailing companies must only offer the service through the use of apps. Therefore, there is a risk that the emerging market will turn into a monopoly. Once the company reaches a monopoly position in the market, there could be inefficiencies and threats to stakeholders. Moreover, as noted by the OECD (2007), a market under monopoly creates political pressure to maintain the protection under regulatory decisions. Therefore, regulators should promote a competitive market to push players to improve and adapt the quality of their service.

Entry conditions in Mexico City were addressed mainly through quality requirements, rather than a quantity regulation. Quality controls were introduced by setting requirements regarding vehicle characteristics, and operation permits for drivers. The practice of quantity regulations was not introduced in an obvious way. The number of cars was not limited, but the annual renewal of an operation permit and the hologram that vehicles must have are regulatory practices that are equivalent to an extreme form of control. As noted above, taxi markets usually experience regulatory capture under this framework. Mexico City did not apply fare regulation or structure. However, it established a limitation in payment options and a 1.5% ride levy for a public mobility fund. It also banned e-hailing companies from receiving payment in cash or prepaid cards. At first, this restriction was not an issue for Uber because payments have only been made by credit card or other electronic payment methods. However, in a developing country, credit-card payment mechanisms may discriminate against the low-income people, who cannot afford this financial benefit. In fact, Uber appealed to a federal judge against this restriction in December 2017. After two years of the approved regulations, Uber realized that cash fares were crucial to conquer the Mexico City market and continue expanding in other Mexican cities. Finally, Uber will pay the 1.5% ride levy, but the fund has not yet been created.

Regarding e-hailing services regulation in Bogotá, the results for Uber and similar companies are not very encouraging. Regulators kept the framework governing traditional taxis and created a new luxury taxi service. Therefore, e-hailing companies must follow the same requirements as traditional taxis. As stated in the case description, taxi regulations are based on exclusivity rights, entry conditions, and economic controls. Similarly, the main aspects addressed in the regulatory framework for e-hailing services in Bogota can be summarized as follows: only available through the use of technological platforms; vehicle restrictions; and fare controls and payment methods restrictions. Regarding exclusivity rights, the government's decree established that vehicles must be enrolled in a taxi company. As mentioned above, the company should request an authorization from the competent authority to operate, meaning that traditional taxis companies are free to provide this new kind of service. Quality regulations were introduced by defining requirements regarding the vehicle age, color, type, and model. There must also be an official holographic band on the side of the vehicle. The decree also bans private vehicles from providing public transportation services. Drivers must meet licensing requirements and courses. There are quantity regulations, even though there are no limits on the number of cars that can be affiliated with a taxi company. As we saw in Colombia, taxi regulations have two mechanisms to incorporate a vehicle into a taxi fleet as a form of entry control. The approval of this decree led to many questions about how the vehicle fleet of this new market will be estimated or controlled, or if the same rules will apply for traditional taxis. Finally, fare controls were implemented in two ways: a fixed minimal fee, and a restriction in the payment method. Therefore, new luxury taxis can only receive payments by electronic means and should charge a legal minimum fee that should be equal to or higher than, traditional taxis.

Comparing the regulations for e-hailing services in Mexico City to those in Colombia, we can see that quite different regulatory approaches were adopted. In the case of Mexico City, regulation for e-hailing companies was as a result of a public participation process, so its effects have been mostly positive. Uber recognized itself as a transportation company and agreed to the parameters established in the regulatory framework. Furthermore, the city set a precedent looking forward to technological and organizational innovations for future challenges. Conversely, Colombia essentially approved the provision of a new taxi service through the use

Table 3.1 Comparison of regulatory approaches for e-hailing services in Mexico City and Colombia

		Exclusivity rights	Entry conditions		Fare regulation
			Quality regulation	Quantity regulation	
Mexico City	Addressed?	–	±	–	–
	Impact	-Monopoly -Vested political in regulatory decisions -System inefficiencies	-Customer security and protection	-No control on environmental and congestion issues -Risk of regulatory capture	-Threat to customer access -Monopoly
Bogota	Addressed?	–	+	–	±
	Impact	-Monopoly -Vested political in regulatory decisions -System inefficiencies	-Customer security and protection -Competency driver, experience	-Risk of regulatory capture	-Equilibrium demand-supply -Prevent gouging by operators

of technological platforms or apps. The decree's long list requirements for vehicles, as well as economic controls, were not a solution for Uber and similar companies, so they are still operating in a "gray zone" (see Table 3.1).

However, we can criticize the lack of consideration of the following points in the regulation approved for e-hailing services in Mexico City. The first point is about the exclusivity rights. According to OECD (2016), a progressive taxation of share market would be a better policy instrument than a fixed quota to prevent TNCs from getting dominant position in the market, as Mexico City did by introducing the 1.5% ride levy. The proportional market share taxation could lead to a cartel, in which the existing license holders could prevent new ones from entering the market. On the other hand, the progressive surcharge could reach punishing rates when the market share gets close to 40 or 45%. Although the 1.5% ride levy could be considered an innovative and fair rule in the taxi market at first glance, this extra amount will affect passengers, who will pay for it, instead of Uber. If we look at other foreign regulation approaches, the ride levy is very low but e-hailing companies must make annual payments to operate. For instance, Uber and similar companies must pay CAN$70,000 a year to the city of Edmonton, Canada, to operate their service, while the ride levy is six cents per ride to finance the cost of implementing the regulation. No compensation has been offered to taxi drivers, despite the predictable reduction in the value of their licenses (Chassin and Msaid 2016).

Second, the quality regulations established should be stronger to address passenger safety and include mechanisms to protect passengers' private data such as names, address, and other relevant information. In cities where violent assault, rape and kidnappings by taxi drivers are still a concern, passenger safety somehow escapes city government control under this regulatory framework. For example, an Uber driver in India was arrested for the rape of a passenger; and others have been accused of stalking passengers (Hanks and Alexander 2014). When regulating e-hailing services, authorities should give priority to maintaining high-security standards and ensuring the safety of passengers.

Third, the absence of fare regulation for e-hailing services has caused anger among passengers. Due to high levels of pollution, Mexico City has established days on which the use of vehicles is restricted, increasing the demand for Uber service. However, because of its surging price patterns, the rates can reach five to nine times the average price. Uber has addressed this situation by limiting the fare to five times its regular price during these days and pushing UberPool services, saying passengers can save up to 40% per trip. However, regulators should be aware of the impact of these policies for taxi users, and establish a fare that allows the efficiencies that e-hailing companies seek to offer, but also guarantee users high-quality transportation services at affordable prices.

With regard to Colombia, regulatory authorities have not envisioned Uber's technology and its innovative business model. Despite the fact authorities have stated that the current regulations exist to protect users, the latter have not been involved in the decision-making process. The number of citizens demanding Uber service keeps increasing. Therefore, it is quite likely that Colombia will be forced to re-evaluate and re-design the approved regulation. As we saw in the introduction, in many cities, UBER kept operating regardless of public authorities regulatory response. This is also the case in Bogota, and conflicts keep arising with the traditional taxi sector and government authorities. Although we do not know with certainty the future of Uber in Colombia, it is hard to believe that a few thousand taxi drivers have more power than millions of users with smartphones and information who also have the right to demand better public service.

We recommend that regulators request e-hailing companies to share their data. Although it is well-known that Uber has refused to open its data, this might change if it becomes a requirement to get the right to operate in a given city. As an example of this, the Transport Authority of the Greater London (TfL 2017), which was the subject of much attention in Fall 2017, given their refusal to renew Uber licenses to operate in London, recently proposed that TNCs will only be granted licenses if they agree on opening their trip data (TfL 2018). When regulating e-hailing services, authorities should take advantage of the emerging technologies, understanding that open, accessible data will allow them to better understand mobility patterns in the city and improve inefficiencies of the public transportation.

Finally, regulators have developed e-hailing regulations based on the tools mentioned in our conceptual framework, but as we saw through the analysis, other considerations are needed to adopt better approaches to regulating e-hailing companies. For instance, the traditional quality regulations did not include protecting of passengers' private data, which is nowadays a concern regarding the customer safety. Likewise, with the emergence of these technology platforms, access to data has become a key rationale for regulating e-hailing services; element that has not been addressed by traditional regulatory frameworks.

3.5 Conclusion

Uber emerged in Mexico City and Bogota, as a competitor for the traditional taxi sector, generating serious conflicts and implications for the regulatory framework that has governed taxi systems. Users have not hesitated to choose Uber rather than traditional taxis, and governments have understood that citizens demand a better service, because that is the characteristic of a competitive market. Regulators must consider how to tighten current regulations, not the other way around, as it unfortunately happened in Bogotá. Although it is still too early to define the future effects of Uber, regulators must think of how the regulatory framework that governs a city's taxi system should be transformed. The regulations should focus on ensuring the quality and efficiency of the offered services and empower the users over the taxi service. Uber has demonstrated its interest in proposing self-driving vehicles, and when this comes to reality (although not in the near future for Latin America), it will again consequently change the transportation landscape. Public authorities also must take this perspective into account when regulating Uber.

Innovation and technology represent an opportunity to improve the efficiency of transportation systems. Regulations should be developed in order to make sure new transport solutions serve the common interest of society. Public leaders must promote the coexistence of diverse, individual transportation systems within cities. Although they are one same industry, TNCs and taxis cannot function under the same rules. Leaders must allow innovation in the menu of transportation service options provided to inhabitants of a given city, under the rules of fair competition. They also should look toward reaching a level playing field, whereby each transportation service could operate efficiently. Likewise, improvements in addressing transportation issues require new models of governance through the participation of stakeholders. In addition, regulators and policymakers should take advantage of the work performed in other cities. These examples would provide authorities new structures that are being developed to achieve better regulations of e-hailing services.

Finally, we believe that more research is needed to clearly understand the impact of e-hailing on transportation cities. Recently, there has been an increasing amount of research trying to show the negative effects of Uber in traffic. For instance, e-hailing has been criticized for increasing congestion (Rayle et al. 2014). Additionally, it was showed that 34% of Uber vehicles on the road were empty of customers in California (SFCTA 2017, quoted in Currie 2018). Therefore, regulation of e-hailing services is not solely about technology, but also about environmental and social effects. In this sense, the effect of Uber has been poorly analyzed. Is Uber an asset by offering new options of mobility, or does it generate more vehicular congestion? Ultimately, there is no city in the world that could improve mobility by adding more vehicular congestion to its streets.

References

Aarhaug J, Skollerud K (2014) Taxi: different solutions in different segments. Transp Res Procedia 14(1):276–283

Audouin M, Neves C (2017) What regulations for ICT-based mobility services in urban transportation systems? The case of ride-booking regulation in Sao-Paulo and Rio de Janeiro. WIT Trans Built Environ 176:95–106

Badger E (2014) Taxi medallions have been the best investment in America for years. Now Uber may be changing that. The Washington Post, 20 June

Barro J (2014) Under pressure from Uber, taxi medallions prices are plummeting. The New York Times, 29 Nov

Beesley ME (1973) Regulation of taxis. Econ J 83(329):150–172

Bowcott O (2017) Uber to face stricter EU regulation after ECJ rules it is transport firm. https://www.theguardian.com/technology/2017/dec/20/uber-european-court-of-justice-ruling-barcelona-taxi-drivers-ecj-eu. Accessed 1 Mar 2018

Buckley C (2015) An examination of taxi apps and public policy regulation. http://clarebuckley.ca/pdf/Clare%20Buckley%20-%20public%20policy%20regulation.pdf. Accessed 16 Nov 2017

Cairns R, Liston-Heyes C (1996) Competition and regulation in the taxi industry. J Public Econ 59 (1):1–15

Cannon S, Summers L (2014) How Uber and the sharing economy can win over regulators. Harvard Business Review, Boston

Chassin Y, Msaid Y (2016) Uber and taxis: Australia opens the door to reform. Montreal Economic Institute. http://www.iedm.org/files/lepoint0216_en.pdf. Accessed 13 Oct 2017

Cohen M, Sundararajan A (2015) Self-regulation and innovation in the peer-to-peer sharing economy. Univ Chic Law Rev Online 82(1)

Corporación Andina de Fomento (CAF) (2011) Desarrollo Urbano y Movilidad en América Latina. Banco de Desarrollo de América Latina, pp 175–188

Currie G (2018) Lies, damned lies, AVs, shared mobility, and urban transit futures. J Public Transp 21(1):19–30

Darbéra R (2015) Principles for the regulation of for-hire road passenger transportation services. Research report. OECD International Transport Forum 2015

DeVany AS (1975) Capacity utilization under alternative regulatory restraints: an analysis of taxi markets. J Polit Econ 83(1):83–94

Douglas GW (1972) Price regulations and optimal service standards. J Transp Econ Policy 6 (2):116–127

Edelman BG (2015) Whither Uber? Competitive dynamics in transportation networks. http://www.benedelman.org/publications/competitive-dynamics-tncs-24nov2015.pdf. Accessed 9 Jan 2018

Edelman BG, Geradin D (2015) Efficiencies and regulatory shortcuts: how should we regulate companies like AirBnB and Uber? Stanf Technol Law Rev 19:293–328

eMarketer (2015) Colombia set to reach 28.6 million internet users in 2015. Accessed 12 Feb 2018

eMarketer (2016) Mobile Colombia 2016: updated forecasts and key growth trends. https://www.emarketer.com/Report/Mobile-Colombia-2016-Updated-Forecasts-Key-GrowtTrends/2001863. Accessed 12 Feb 2018

Farren M, Koopman C, Mitchell M (2016) Rethinking taxi regulations: the case for fundamental reform. Mercatus Research, Mercatus Center at George Mason University, Arlington, VA

García C (2016) La Regulación de Uber en la Ciudad de México, la ganancia de los consumidores y el problema público de la movilidad. Lat Am Iberian J Law Econ 2(2):39–63

Geloso V, Guénette J (2014) Ride-sharing applications and the future of urban transportation. Montreal Economic Institute. https://www.academia.edu/9787693/Ride-Sharing_Applications_and_the_Future_of_Urban_Transportation. Accessed 20 Oct 2017

Geradin D (2015) Should Uber be allowed to compete in Europe? And if so how? George Mason University, Law & Economics Research Paper Series No. 15–29

Grand E, Khosla S (2015) Here's everywhere Uber is banned around the world. Business Insider. http://www.businessinsider.com/heres-everywhere-uber-is-banned-around-the-world-2015-4. Accessed 12 Feb 2018

GSMA (2016) Country overview: Mexico. https://www.gsmaintelligence.com/research/?file=44866ee04f5cc721e249569adbd505f7&download. Accessed 9 Jan 2018

Hanks E, Alexander S (2014) Same industry, same rules? When rideshare comes to town. https://casesimportal.newark.rutgers.edu/sites/default/files/files/703_Hanks_When%20rideshare%20comes%20to%20town.pdf. Accessed 20 Nov 2017

Harding S, Kandlikar M, Gulati S (2016) Taxi apps, regulation and the market for taxi journeys. Transp Res Part A Policy Pract 88:15–25

Harrington D (2012) E-commerce in Latin-American: the success of MercadoLibre. http://davidwarrenharrington.weebly.com/uploads/4/7/0/8/47082859/e-commerce_in_latin_america_mercadolibre.pdf. Accessed 23 Jan 2018

Harris L (2002) Taxicab economics: the freedom to contract for a ride. Georgetown J Law Public Policy 1:195–222

Ibañéz M (2012) Viabilidad Técnica y Financiera del Sistema de Taxis en el Sistema Integrado de Transporte. http://www.bdigital.unal.edu.co/8596/1/300457.2012.pdf. Accessed 20 Nov 2017

Islas V, Hernandez S, Arroyo J, Lelis M, Ruvalcaba JI (2011) Implementing sustainable urban travel policies in Mexico. International Transport Forum Discussion Paper No. 2011/2014

Li HR (2016) Taxi positioning in the new age of internet and industrial development research. Procedia Eng 137:811–816

Lopez M (2012) El Transporte de Pasajeros y el sistema vial EN la ciudad de México. Biblioteca Jurídica Virtual del Instituto de Investigaciones Jurídicas de la UNAM, México. http://biblio.juridicas.unam.mx/libros/6/2735/12.pdf Accessed 1 Dec 2017

Moon Y (2015) Uber: changing the way the world moves. Harvard Business School, Case 101

Nelson/Nygaard Consulting Associates (2008) Santa Monica taxi study—technical memorandum. http://nelsonnygaard.com/wp-content/uploads/2008/03/SANTA-MONICA-Taxi-Study.pdf. Accessed 20 Feb 2018

Newcomer E (2016) Can Uber conquer Latin America? Bloomberg. https://www.bloomberg.com/news/articles/2016-10-13/can-uber-conquer-latin-america. Accessed 20 Nov 2017

OECD (2007) Taxi services: competition and regulation 2007. http://www.oecd.org/regreform/sectors/41472612.pdf. Accessed 12 Feb 2018

OECD (2016) App-based ride and taxi principles for regulation. https://www.itf-oecd.org/sites/default/files/docs/app-ride-taxi-regulation.pdf. Accessed 20 Feb 2018

Opina Bogotá (2017) Uber vs Taxis: entrevista al Secretario de Movilidad de Bogotá. https://www.youtube.com/watch?v=K8OwW6ywO38. Accessed 22 Nov 2017

Parametria (2013) Movilidad y transporte en el Distrito Federal. Investigación Estratégica Análisis de Opinión y Mercado. http://www.parametria.com.mx/DetalleEstudio.php?E=4539#.VwwcPVt4oqM.email. Accessed 9 Nov 2017

Rayle L, Shaheen S, Chan N, Dai D, Cervero R (2014) App-based, on-demand ride services: comparing taxi and ridesourcing trips and user characteristics in San Francisco. University of California Transportation Center. https://www.its.dot.gov/itspac/Dec2014/RidesourcingWhitePaper_Nov2014.pdf. Accessed 20 Nov 2017

Rogers B (2015) The social costs of Uber. Temple University Legal Studies Research Paper No. 2015-28

Salanova J, Estarda M, Aifadopoulou G, Mitsakis E (2011) A review of the modeling of taxi services. Procedia Soc Behav Sci 20:150–161

Sanchez L, Avendaño S, Coronel Y, Castellanos L (2016) Uber an innovative deregulated business or an infringer of the free competition? https://www.competitionpolicyinternational.com/uber-an-innovative-deregulated-business-or-an-infringer-of-the-free-competition/. Accessed 10 Feb 2018

Schaller B (2007) Entry controls in taxi regulations: implications of US and Canadian experience for taxi regulation and deregulation. Transp Policy 14(6):490–506

SFCTA (2017) TNCs today: a profile of San Francisco transportation network company activity. http://www.sfcta.org/sites/default/files/content/Planning/TNCs/TNCs_Today_112917.pdf. Accessed 4 Mar 2018

Silverstein S (2014) These animated charts tell you everything about Uber prices in 21 cities. Business Insider. http://www.businessinsider.com/uber-vs-taxi-pricing-by-city-2014-10. Accessed 1 Mar 2018

Statista (2018) Numbers of smartphone users in Mexico from 2015 to 2022 (in millions). https://www.statista.com/statistics/270970/number-of-smartphone-users-mexico/. Accessed 10 Feb 2018

Strong C (2015) When apps pollute: regulating transportation network companies to maximize environmental benefits. Univ Colo Law Rev 86

TfL (2017) Licensing decision on Uber London Limited. https://tfl.gov.uk/info-for/media/press-releases/2017/september/licensing-decision-on-uber-london-limited. Accessed 20 Feb 2018

TfL (2018). London taxi and private hire. http://content.tfl.gov.uk/private-hire-policy-statement.pdf. Accessed 20 Feb 2018

Torres N (2016) Nearly a year on, Mexico City Uber regulation stuck in neutral. https://www.thepeninsulaqatar.com/article/17/06/2016/Nearly-a-year-on,-Mexico-City-Uber-regulation-stuck-in-neutral. Accessed 3 Nov 2017

Uber Estimate (2018) Uber cities. https://uberestimator.com/cities Accessed 22 Nov 2017

United Nations (2014) The word's cities in 2016. http://www.un.org/en/development/desa/population/publications/pdf/urbanization/the_worlds_cities_in_2016_data_booklet.pdf. Accessed 10 Jan 2018

WSJ (2014) Uber: the state of play. The Wall Street Journal, 15 Dec

Yin RK (2009) Case study research: design and methods (4th edition). Can J Action Res 14(1):69–71

Yussoff RM, Karim NA, Daud A (2015) Impact of satellite-based dispatch systems for taxi services in the urban areas: a literature review. In: Global Conference on Economics and Management Science 2015

Maria Lorena Puche holds a Bachelor of Civil Engineering from UNEFA, Maracay (Venezuela), and a Master's Degree in Innovative Governance of Large Urban Systems (IGLUS) from EPFL, Switzerland. She has over seven years of experience in city management and public administration. Her research interests are based on her passion for the governance of transportation systems, new regulatory approaches, and governance models for emerging innovative solutions in the transport sector as well as innovative concepts such as Mobility-as-a-Service (MaaS).

Chapter 4
Governing Carsharing as a Commercial or a Public Service? A Comparison Between France and Japan

Bruno Faivre d'Arcier and Yveline Lecler

Abstract Based on case studies in Japan and France, this chapter analyzes the conditions needed for station-based carsharing to take off. Using the Japanese perception of transportation as a commercial service, and the French perception of it as a public service, this chapter shows that round-trip and one-way services are two different markets. These systems, in which use, operating costs, vehicles, and impact on car dependency are not the same, developed quite differently in each country. Although commercial round-trip services have grown fast in Japan, France relied more on electric vehicles one-way services supported by local public authorities. Social, institutional and regulatory contexts, and user demand explain the differences, but the roles of private and public actors also matter.

Keywords Carsharing · Round-trip · One-way · France · Japan

4.1 Introduction

The concept of carsharing first appeared in the 1950s as a means to avoid high costs of car ownership (Communauto, n.d.; Shaheen and Cohen 2007). In the late 1960s, Friedman (1972) identified advantages of sharing the use of a car in terms of reducing traffic congestion, parking space needs, travel costs and air pollution in peak periods, and promoting public transportation (PT) in regard to public or societal issues. This still appears relevant today. Feedback from the first experiments showed that carsharing services encouraged individuals to change their travel habits and reduce driven mileage (Shaheen and Cohen 2013).

B. Faivre d'Arcier (✉)
Transport Urban Planning Economics Laboratory, University of Lyon: Lyon 2, Lyon, France
e-mail: bruno.faivre-darcier@laet.ish-lyon.cnrs.fr

Y. Lecler
Institute of East Asian Studies, University of Lyon, Sciences Po Lyon, Lyon, France
e-mail: yveline.lecler@ens-lyon.fr

M. Finger and M. Audouin (eds.), *The Governance of Smart Transportation Systems*,
The Urban Book Series, https://doi.org/10.1007/978-3-319-96526-0_4

The idea of shared goods, rather than individual ownership, spread worldwide in the 2000s. Sharing cars, which are parked 23 h a day, became an even more relevant choice, opening new business opportunities. Studying Carlink in the US, Shaheen (1999) stated that carsharing can "*reduce traffic congestion, air pollution, and government spending*," while the broad development of information and communication technologies (ICTs) made carsharing less inconvenient and smart thanks to intelligent registration and reservation systems. This might explain why apart from many environmental associations which were and still are promoting carsharing, diverse companies are entering the sector. This might also have contributed to renewed interest of local and national public bodies in supporting carsharing as a more sustainable mode of mobility.

Based on empirical observation and interviews, this chapter aims to analyze the development of station-based carsharing in France and in Japan, which both came into the carsharing business relatively late. Carsharing is defined as a short-duration rent-a-car system, through membership and pay-as-you-go pricing (including insurance, fueling, maintenance, and cleaning). It differs from car rental, which needs a contract for each rental and a face-to-face contact at an office. It also differs from carpooling, in which individuals share the same car for the same trip, thanks to platform managers who just organize matching people. Carsharing can be station-based, in which cars are picked up and returned at on- or off-street stations. It can also be free-floating, in which cars can be picked up and returned anywhere in the city. While free-floating carsharing is a one-way service, station-based carsharing can be either round-trip, when cars have to be returned at the station they were picked-up, or one-way when cars can be returned at a station different from picked-up one. Given that free-floating carsharing is just starting in France and does not yet exist in Japan, this study only focuses on station-based carsharing.

In the recent years, Japan experienced a huge increase in membership, whereas the number of users in France stagnated. Therefore, comparing France and Japan is interesting, although station-based carsharing still remains niche markets in both countries. The development of round-trip and one-way systems is quite different in each country, due to differences in societal, policy, and regulatory contexts. For historical reasons, Japan considers passenger transportation a commercial activity, mainly relying on private companies for implementation, funding, and operation. France mainly views it as a public service, organized and partially funded by a public authority, which often entrusts a private company to operate it. Observing the respective evolution of round-trip and one-way markets in France and Japan, this chapter intends to address the following question: *Should station-based carsharing develop as a commercial or a public service, and what are the conditions required for it to take off?*

This chapter is organized as follows: Part 2 will briefly present the literature on carsharing and the theoretical framework relevant to analyze its development, before introducing its evolution in France and Japan. Parts 3 and 4 empirically discuss what happened in each country. Part 5 analyzes the reasons for differences between the countries, as well as their possible impact for future prospects. In Part 6, the conclusion will briefly summarize findings. Unless specifically referenced, all the empirical data and information was collected through interviews (listed in the references section) conducted with different actors within the carsharing sector in France and Japan.[1]

4.2 State of the Art and Carsharing Evolution in France and Japan

Inspired by the sharing economy and the development of ICT, the literature on carsharing is abundant. Based on surveys conducted in numerous countries, it mainly questions the emergence of this new mobility system. On the one hand, its characteristics are analyzed using traditional, transportation study approaches, through users' profiles (Loose 2010; 6t-bureau-de-recherche 2016), or type and frequency of uses (Cervero et al. 2007) to estimate the potential market and its impact in terms of complementing or substituting for other modes. On the other hand, some studies are more concerned with the sustainability or smartness of the system, so demonstrate how carsharing might reduce car ownership, congestion, or other car-related negative externalities (Martin and Shaheen 2011). Finally, other studies are concerned by the market potential and its associated business models (Steininger et al. 1996; Shaheen and Chan 2015; Boston Consulting Group 2016; Shaheen and Cohen 2016).

In Japan, academic literature on carsharing is less important. Survey reports of the Ministry of Land, Infrastructure, Transport and Tourism (MLIT) or consultancies provide most of the data on experiments. Some scholars are studying the best location for stations through demand simulation (Nakamura et al. 2017), while others are looking at stakeholder strategies involved in urban transportation or carsharing to see if there is cooperation (Kato et al. 2015). However, most papers focus on e-carsharing, with special attention on micro-mobility e-carsharing (Mizokami et al. 2015).

The transition toward sustainable or smart mobility is often discussed from a socio-technical system and multi-level perspective (MLP) approach (Geels 2012). Such papers look at the conditions for the emergence of a new product (such as electric vehicles—EVs) or service (such as navigation systems) to see how these innovation niches can generate change at the regime level, and/or how the landscape level may impact or be impacted. Although the evolution of carsharing in France and Japan could be explained by changes in technology, socioeconomic norms, and regulation, such an approach probably underestimates the role of the demand side. At the opposite end, transition can also be seen through behavioral studies in terms of modal choice and incentive (information, experimentation, or price signal) to encourage each individual to choose more sustainable or smart behaviors. However, as past experiments or incentives have shown, it does not

[1]This study benefited from a two-month invitation to Kansai University during the fall of 2017.

work that easily. Behaviors are the observable expression of social phenomena (Spurling et al. 2013), rooted in cultural values and world vision, previous learning and routines, and former policies and institutions. This means that access to resources, new technologies or services is important, but that history and path dependencies also matter. The theories of practice, which put practices (action of doing) instead of individual behavior or technological systems at the core of the analysis, seem better able to take into account the different dimensions of the issue. Practices are defined (Spurling et al. 2013) as socially recognized activities in which people engage in consuming resources to accomplish actions (practices as performance), such as bathing, skiing, driving, or cycling. According to Watson (2012: 492): "*Systems persist and are transformed only through the flow of practices—of action and doing—which comprise them.*"

The approach through practices (as entities) that "*integrates both behaviors and their material, social and cultural contexts*" (Spurling et al. 2013: 19), but also addresses the systemic change dimension (systems of practices), can better grasp the conditions for services' attractiveness, whether round-trip or one-way, and the innovations or interventions to support their takeoff or sustainability. Indeed, technology matters, but as it will be shown in this chapter, the actions of actors throughout the system explain the different development and governance of carsharing systems in France and Japan.

Due to climate change, energy transition, or congestion in cities, policies since the 1990s tend to intervene in the car driving practice, experimenting to make mobility more sustainable. Carsharing, which relies on the same elements as car driving, in terms of material (vehicle and road infrastructure), competence (driving), and meaning (mobility), occupies a singular position. If "*re-crafting practices focuses on making driving less resource intensive, but does not seek to change patterns and volumes of private car use*" (Spurling et al. 2013: 27), then carsharing must be considered a new practice aimed at substituting car ownership. In that sense, it competes with (private) car driving. However, similar to car driving, it also competes with alternative mobility solutions. As practice theories show (Spurling et al. 2013; Watson 2012), a practice needs defection from other practices to develop. In promoting carsharing, policy-makers expect to recruit members among car owners, abandoning car purchasing or at least not using private cars under certain circumstances or at certain times.

During the 1990s, carsharing started in France and Japan, developing through local initiatives from individuals, but also through several projects from governmental support. Linked to the development of EVs, these projects were often launched under the leadership of carmakers. Building on progress made in the ICT field, many applications for real-time information (booking, vehicle location and follow-up, pricing, and access cards) were then available to help operate such systems. These experiments, as a test bed for developing EVs and demonstrating ICT technological components, lasted only a couple of years in France (TULIP Project by PSA, PRAXITELE Project from Renault) and Japan (Honda ICVS, Toyota Motor Company [TMC] Crayon System). Grants were provided by the government, but could not recruit enough practitioners, so it appeared impossible to continue once the grants ended. These experiments were also a learning process for public authorities and companies, in that they showed the difficulties to overcome for the market to take off. Until the mid- to late-2000s, France and Japan lagged behind other countries, when looking at the number of members compared to their population.

Statistics on carsharing membership are difficult to compare. In Japan, an annual survey by the Foundation for Promoting Personal Mobility and Ecological Transportation (Eco-Mo Foundation) gives, since 2002, a precise idea of the carsharing evolution, in terms of members, stations, vehicles, and carsharing service operators.[2] In these surveys, individual and corporate members are all registered drivers. In France, there is no annual survey. Data about carsharing membership comes from a research and consultancy office (6t-bureau-de-recherche 2017) that gives numbers in terms of registered members or active members (members renting a car at least once a month). Considering that approximately 50% of members were active in several of the carsharing systems according to the definition given by 6t-bureau-de-recherche, numbers for France were estimated from that ratio.

Although these numbers must be taken cautiously, the comparison clearly shows different evolutions. Between 2006 (the first year of available data for France) and 2010, the number of members continuously increased in both countries. However, France had 3500 members in 2006 and 28,000 in 2010, while Japan had 1712 and 15,894 during the same time period. Since 2010/2011, the number of members dramatically increased, reaching a much higher level in Japan than in France. There were nearly 200,000 members in France in 2016, and 846,240 members in Japan (1,085,922 by 2017).

The social context and user demand partially explain these different evolutions, but as case studies will show, the types of intervention by public and private actors were the first drivers for change in both countries, although leading to quite different results.

4.3 Carsharing in France

As the concept of sustainable development has been imported within French legislation, the government has begun promoting carsharing, thanks to the national energy agency (ADEME) which financially supported the first initiatives (APUR 2008). Since transport regulations are based on the notion of public service, carsharing has been considered as a form of rental service. This made it impossible for municipalities to promote carsharing by providing financial grants or reserved on-street parking spaces, which is required to make the service visible to people. It did not prevent some initiatives to be taken. In 2007 for example, Paris decided to create a label (APUR 2008) specifying the level and quality of carsharing service:

[2]Available in Japanese at http://www.ecomo.or.jp/environment/carshare/carshare_top.html.

access to stations, type of vehicles, availability of cars, and the structure of fares. The label also gave the operator an obligation to transmit all information on users and usage to the municipality. In return, the city offered some advantages to the operators: a 20% reduction on paid parking, reservation of on-street parking spaces, and communication support. However, apart from such specific cases, the legal framework hampered larger development of carsharing.

This is why local officials put pressure on the government to modify the legislation. In 2010, the Law on the Environment (Law No. 2010-788 July 12, 2010, art. 54) created a national label legalizing reserving on-street parking spaces. Citing the use of parking spaces reserved for disabled people on public streets, the law referred to the notion of public utility to justify this legalization. Moreover, the last law on the modernization of public action (Law No. 2014-58 of January 27, 2014) transformed the Public Transport Authorities (PTA) into the Public Mobility Authorities (PMA), giving the new structure extended competences to all modes of transportation. Such a PMA can take the responsibility of creating a carsharing public service, in case of private initiative failure. A part of the revenue of the local transportation tax can then be used to fund investment and operation, as is the case for any PT.

4.3.1 Round-Trip Carsharing in France: Stagnant Demand

Round-trip started in Paris, in the late 1990s with Caisse Commune (which became Mobizen, and was bought by Communauto in 2012), in Strasbourg with Auto'trement, and in La Rochelle under the Liselec project with EVs (now Yélomobile). The number of services quickly extended to 19 in 2008 (CERTU 2008), and to 31 in 2016 (6t-bureau-de-recherche 2017). However, after rapid growth during the first few years, the number of users seems to have reached its limit, at less than 40,000 members, for 1846 vehicles, and 780 stations throughout France (6t-bureau-de-recherche 2017).

From 2002, operators outside Paris started collaborating through the Company France Auto Partage (FAP), which had 10 operators (85 stations, 177 vehicles, and nearly 3000 members) in 2008. This was an opportunity to exchange experiences and know-how between very different operators (associations, cooperatives, and semi-public companies). Moreover, some resources were pooled, such as management software, a call center, and a central purchasing body for vehicles (APUR 2008). In 2013, this cooperative network was renamed Citiz and in 2017 had 15 operators in 50 French cities (300 stations, 750 cars, and 16,000 members; *LPA Magazine* 2017) including Lyon Parc Auto service, used as an illustration below.

Citiz, Lyon Parc Auto (LPA)

The first carsharing system in the Grand Lyon Urban Community was created in 2003 as an association of volunteers (*La Voiture Autrement*). As such, it temporarily benefited from small, indirect, financial support from the public authority. Due to financial difficulties, LPA, a semi-public company in charge of managing public parking, acquired the service (24 cars) in 2007 and progressively increased the fleet to 107 cars (January 2018). At the beginning, stations were limited to a dozen of underground parking buildings managed by LPA. However, thanks to legislation change, since the early 2010s, small on-street stations (two to three cars) were developed, mainly in the central area. The service now has 42 stations, 24 of which are on-street (January 2018).

After a relatively rapid increase in membership during the first few years (Fig. 4.1), the number of users has stabilized or even slightly decreased, due to the launch of competing services such as Bluely. Since 2015, membership is growing again to reach approximately 2500 drivers, thanks to a communication campaign and more diversified, attractive tariffs. The average rental duration remains stable at 7–8 h, for a distance of 65–70 km per trip, with respective medians of four to four-and-a-half hours and 20 km. Longer-duration rentals are for weekends. Corporate membership is progressively increasing and represents 27% of the total (2017). The small size of the service is one of the reasons for its lack of a business model, but the turnover increase and the financial improvement suggest profitability could be achieved within a few years.

The service is mainly used for visits and leisure (particularly outside the city), or carrying goods. The average occupancy of vehicles is 2.1 people, which is 65% more than for private cars. While only 39% of subscribers did not own a car before joining LPA carsharing service, 41% abandoned car ownership after becoming member. In total, 80% of LPA carsharing users do not own a car. This reduction of car use in favor of walking, cycling, and PT not only concerns private car use, but also car rental and taxis.

4.3.2 The Attractiveness of EVs' One-Way Carsharing Systems

There are still only a few one-way carsharing systems in France. Outside of the Paris region with Autolib, EVs one-way carsharing is slowly developing. The city of Nice implemented the Auto Bleue service in 2011 (140 cars, 68 stations, and 2500 active members in 2016), while the Bolloré Group developed its service in 2014 in Bordeaux (Bluecub: 200 vehicles, 80 stations) and Lyon (Bluely: 250 vehicles, 100 stations) at its own initiative. For example, Bluely has no public

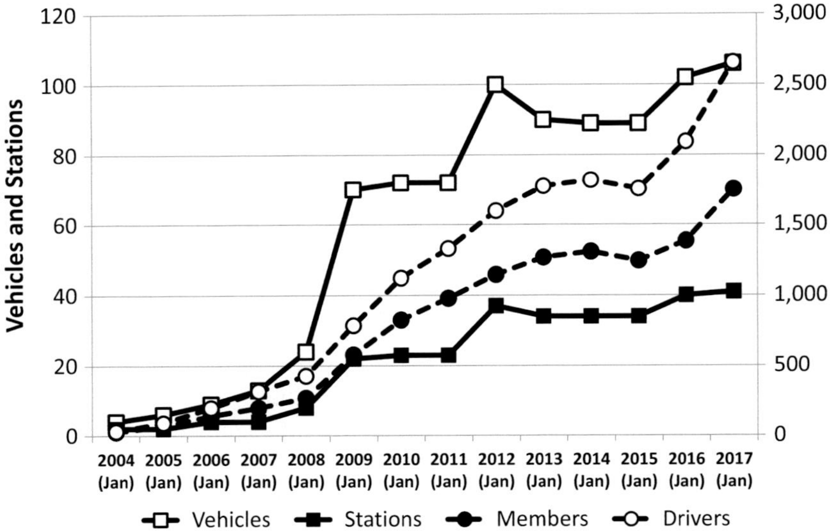

Fig. 4.1 Lyon LPA Citiz service main characteristics. *Source* Based on data from LPA unpublished document, 2018

financial support in Lyon. However, as a service labeled by the Grand Lyon Métropole, it benefits from a contract to rent on-street parking spaces.

Autolib

Imagined as a technological showcase for a new mobility service aimed at complementing the PT network, this EV one-way system came from a group of 19 municipalities, including Paris, which created the Autolib Intercommunal Association. In 2013, it became the Autolib Metropole, with about 30 municipalities. In 2016, after integrating management of the Velib bike-sharing system, it was renamed Autolib Velib Métropole. A call for bids was issued in 2009, suggesting large development over a wide territory. Due to the size of the expected service and the need for charging stations, the municipalities thought that a private initiative would fail. Therefore, they decided to develop it as a public service under the responsibility of public authorities, but entrusted a private company with operation through a Public-Service Delegation Contract.

Several consortiums of companies involved in transportation services made bids. Bolloré, an industrial group specializing in high-capacity storage systems, was interested in using carsharing as a showroom for its Bluecar and, more precisely, its battery technology. It won the bid, created a subsidiary (Autolib Company), and started operating with 250 cars and 250 stations at the end of 2011. It had a final objective of 3000 cars and 1000 stations.

Today, the Autolib Velib Metropole association covers 98 municipalities (665 km^2) and generates 5.76 million rentals per year. This is partly due to the extension in the service area, which increased the number of vehicles to almost 4000 (33% higher than the initial objective, see Fig. 4.2). Obviously, stations located far from the city center do not work as well as those in the city of Paris, leading to financial difficulties.

According to the technical specifications of the call for bids, the investment was estimated at €200 million, with an annual operating cost of €80 million and a turnover of €95 million. The city of Paris committed to contribute €35 million, and the other municipalities agreed to pay €50,000 per station. Considering that profitability would not be achieved before the number of members reached 200,000 for 3000 cars because of the high cost of EVs, the 12-year contract provided that the public would cover cumulative losses exceeding €60 million, attributable to Bolloré. Over the past two years, Autolib had negative gross operating profit of more than €20 million each year, and the public authority is concerned about the risk of bearing a high financial burden in the coming years. Some municipalities are even considering leaving the system.

Annual ticket individual holders grew from 18,775 in 2012 to 40,974 in 2013, and to 111,331 in 2016, whereas there are still only a few corporate members (pro and Utilib), at 2443 in 2016 (Autolib Métropole 2017). Monthly, weekly, and daily tickets are also available, but sales are decreasing over time for the benefit of annual tickets. A survey conducted in 2014 (6t-bureau-de-recherche 2014) shows that one-way user profiles were similar to those for round-trip users, but 57% of one-way users accessed the service more than twice a week. The average rental lasted 40 min, for a distance of only 9 km. Most trips were done within Paris; 62% of members used the service for trips related to work (32% regularly); 66% considered Autolib more convenient than private cars, due to reserved parking spaces; and 25% stated it was more convenient than PT. Substitution also concerned motorcycles or taxis. One-way service replaced on average, compared to private car use, three cars and two parking spaces. However, overall mileage was only reduced by 11%, which is much lower than round-trip systems.

Even if part of the difficulties might be attributed to the public-service design, leading to a lack of supply optimization, this does not explain why the number of one-year season tickets is now slightly decreasing. This seems partly due to lower user satisfaction (notably due to lack of cars' cleanliness) and that carsharing companies are now competing with Transportation Network Companies (such as Uber).[3]

[3]http://www.lepoint.fr/automobile/autolib-en-perte-de-vitesse-14-11-2017-2172121_646.php. Accessed 6 February 2018.

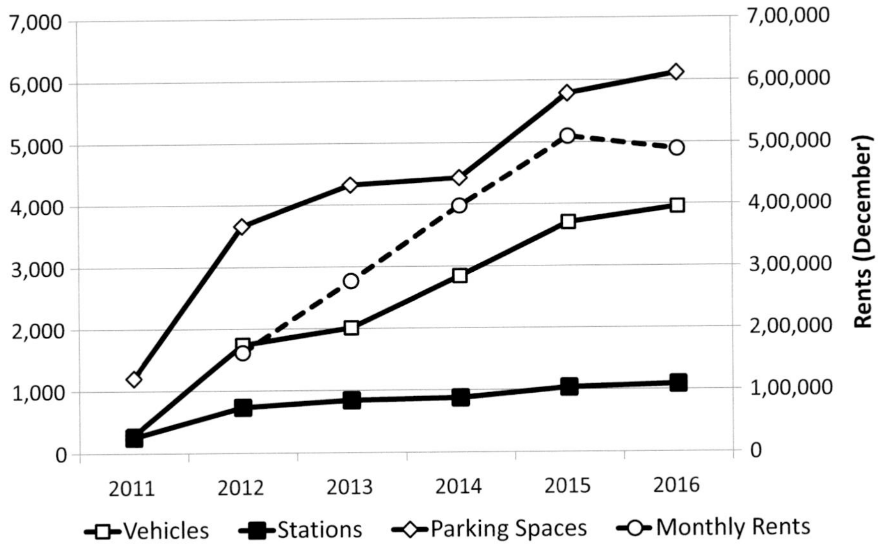

Fig. 4.2 Autolib service main characteristics. *Source* Based on data from Autolib Métropole 2016 activity report

4.4 Carsharing in Japan

The carsharing business is classified as a car-rental business in Japan, requiring a car-rental license. This is why companies other than car-rental businesses could not easily offer such services, or had to do so in association with a car-rental company. This was the case for Park 24, which started the business after integrating Mazda Rental. In addition, all vehicles in Japan must be preregistered with a fixed parking slot, as on-street parking is not authorized, as stipulated in the Road Transport Vehicle Act (Act No. 185 of 1951, Article 7) and the Parking Lot Act (Act No. 106 of 1957, Article 3). This applies to individuals, companies, and car-rental, and carsharing businesses. The law also requires that car-rental companies carry out all matters (lending, vehicle preparation/maintenance, driving license checks, and payments) face-to-face at an office. Moreover, this office must be located no more than 2 km from the rental vehicle's pre-registered fixed parking slot. This has long been a real barrier to the development of carsharing in Japan, partially explaining its late takeoff.

However, these provisions eased from 2004 to 2006, first in relation to the Special Zones for Structural Reform law (Cabinet Office 2003), promulgated in December 2002 (launched in April 2003), and then extended to all of Japan (2006). Through the MLIT enforcement rule,[4] the carsharing type of rental (out of a

[4]See www.mlit.go.jp/report/press/jidosha03_hh_000176.html (in Japanese).

car-rental business) can accurately handle lending vehicles by utilizing ICT. Accordingly, if all operations are done through ICT, carsharing can be exempted from face-to-face management. The unmanned off-road parking slot is then considered an office where the car is registered, fulfilling the prerequisite of the pre-registered fixed parking slot.

These changes did not abolish the 2-km distance requirement between the office and the parking slot. So while easing the development of round-trip systems, this was not the case for one-way systems, in which the car was not returned to the same "office." Therefore, MLIT issued an official notification in 2014,[5] making clear that the off-road parking slot to which the car was returned was its office, even if it differed from its pick-up slot. This aimed to help one-way services take off, especially those using EVs.

4.4.1 Round-Trip in Japan: A Market Dominated by One Company

Like in France, some Japanese NPO entered round-trip carsharing early, but on a very small scale. What differs from the French case is the involvement of parking management companies since the early 2000s. This is due to a very different context in terms of transportation and mobility. The first element comes from the regulation described above, which means that paid parking is the rule. This is why parking management is an important sector in Japan. This sector involves big companies that have often also developed car rental as side business like most carmakers have done. Real estate companies are another category of actors in the sector, which are also absent in the French case. These companies include parking lots in their condominium construction. However, because of space or environmental issue (and probably both jointly), such companies recently shifted from numerous parking slots to offering carsharing. This is the case for Mitsui Fudosan, the third largest carsharing provider in the Japanese market. Another difference lies in the fact that PT networks (private and public railways) are widely developed in Japan, especially in the three main urban areas. A large majority of home-to-work trips are done with PT. Private cars are not used for commuting or everyday activities. This means that carsharing seems well adapted to an urban population whose need for cars is limited to unusual or special trips.

In 2002, Orix Auto Lease Corporation was the first to start a carsharing system in the framework of a city of Yokohama pilot project. In 2005, the company became Orix Auto Corporation by integrating seven companies involved in car leasing and rental activities. The system that the company started, which progressively transformed into a commercial service in 2005, was first proposed with EVs.

---

[5]Issued March 2014, entering into force in September, See www.mlit.go.jp/report/press/jidosha03_hh_000176.html (in Japanese).

However, people did not rent them because they were afraid of the lack of range and the charging duration. Over time, the electric fleet disappeared in favor of gasoline cars.

As we have seen, the size of the carsharing market has drastically increased since 2011. Eco-Mo Foundation (2017) identified approximately 30 carsharing providers in Japan, but only six had more than 10,000 members at the end of 2016. The offensive strategy of the Park 24 group explains this takeoff.

Times Car Plus: Park 24 Group's Carsharing Service

Park 24 started its activity in the late 2000s, some years before carsharing became fashionable. With the integration of Mazda Car-Rental to the group, commercial activity began in 2009 (1030 vehicles and 746 stations in October 2010), with a strategy to rapidly increase the fleet. After several reorganizations, Times 24 Co. was created in 2010 to take charge of the carsharing service, renamed Times Car Plus (TCP) in 2014. Mazda Car-Rental was renamed Times Mobility Networks, while the parking business was transferred to Times 24 in 2011.

Times 24 developed an original strategy, giving it a competitive advantage over other parking-management companies. The group has its own underground or elevated parking buildings, but it also opened small street-level car parks by continuously looking for vacant land parcels within cities. These small car parks, in which one or two spaces could easily be reserved for carsharing, can be considered as on-street stations. The vacant parcels are leased under two-year renewable contract with their owners. Investment is low, and if one parking lot closes, another can be opened since vacant land parcels are always available.

As Fig. 4.3 shows, TCP totaled more than 20,000 cars, more than 10,000 stations, and more than 900,000 members by October 2017. A 2016 company survey of 5616 respondents showed that the average frequency of use was two to three times per month for 3.75 h, to travel 40 km. Shopping was the main purpose for trips (70%), but also for pleasure driving (35%) or carrying goods or people (33%). Fourteen percent of members owned a car, while 53 had given up their car. As fares are fixed to be close to PT prices, such services attract many young Japanese drivers who don't wish owning a car, or are not yet able to buy one. User satisfaction came from the possibility to use a car only when necessary (79%), not paying for fuel or parking (66%), the proximity of stations (54%), and the 24/7 availability of the service (52%).

With a 70% share, TCP dominates the market. Its main competitors Orix Auto Corporation (170,000 members; 2600 vehicles; 1531 stations) and Mitsui Fudosan Realty (Careco: 57,000 members; 1760 vehicles; 1159

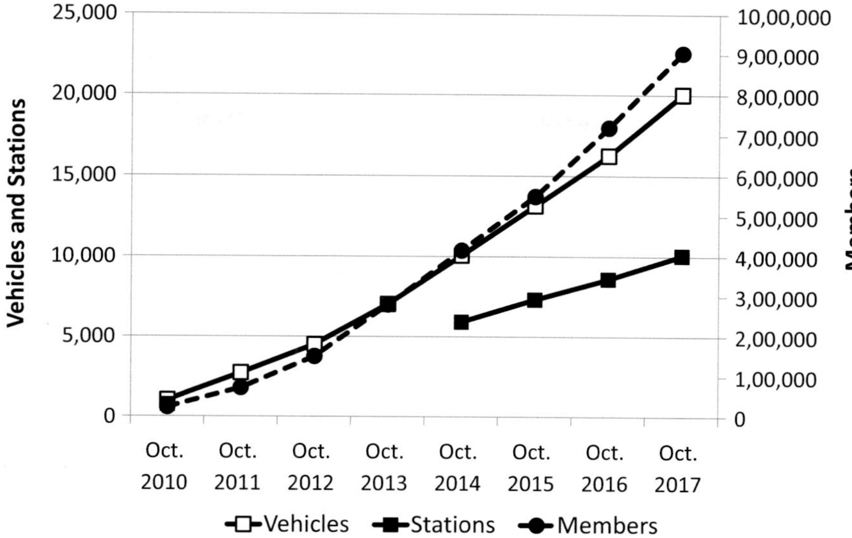

Fig. 4.3 Development of Time Car Plus service. *Source* Company data

stations) remain far behind. TCP was first mainly developed in the three largest urban areas, and then expanded to all of Japan. It was particularly established at each Shinkansen Railway station, as a last-mile service for companies, which represent 35% of members. TCP's quick expansion was facilitated by the financial capacity of the Park 24 Group, which could easily invest in cars and stations, as it was highly profitable in its parking-management activity. By 2014, the carsharing system generated profits for the company, which engaged in a long-term strategy (Fig. 4.4).

However, Times 24's strategy must be linked to other elements to explain its success. Apart from the looser regulations in 2004–2006, another element happened as a consequence of the 2008 financial crisis. Companies tried to cut costs, and discovered that carsharing could be a better solution than increasing or keeping a fleet of company cars. Another element came from the younger generation's attitude toward car ownership, which is no longer considered a visible sign of social success. Instead, as Bardhi and Eckhardt (2012) have showed for Zipcar, carsharing gives a smarter image to users.

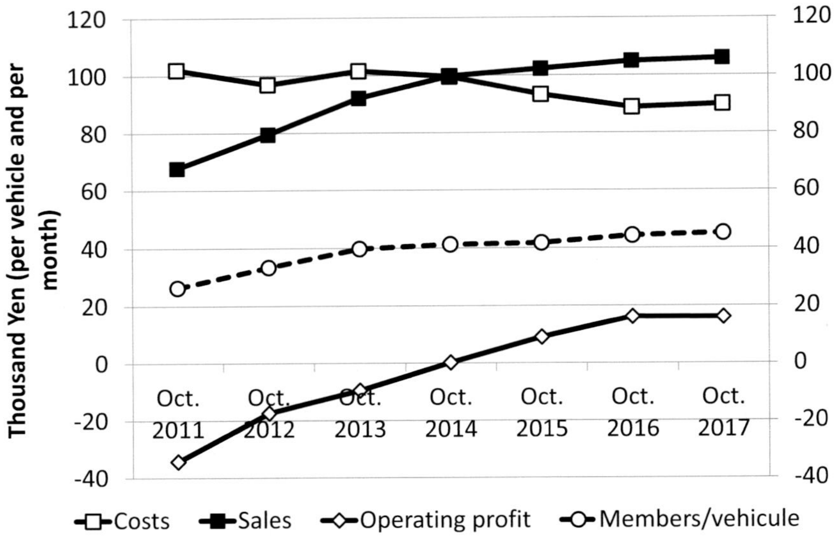

Fig. 4.4 Financial results of Time Car Plus service. *Source* Company data

4.4.2 Japan: Experimenting with EVs One-Way Carsharing Systems

Japanese carsharing companies were not reluctant to introduce EVs into their fleet, but they seem very cautious about one-way systems. Therefore, public-policy measures have pushed for EV one-way carsharing in a two-part, specific context.

First, the Ministry of Economy, Trade and Industry (METI) launched the Next-Generation Energy and Social System Program (Faivre d'Arcier and Lecler 2014) in 2010 to experiment with smart-grid technologies, promote renewable energies, and further address the question of lifestyle changes. The issue of transportation and mobility was one of the elements taken into account in so-called "demonstrators of smart communities." Among the four selected projects, two tested one-way e-carsharing in relation to energy consumption and lifestyle changes: Ha:mo (Toyota city) and Choi Mobi (Yokohama). Second, both services used micro-vehicles and benefited from MLIT subsidies under a program opened in 2012 to experiment with micro-e-mobility. Surveys showed that there were no passengers for 75–80% of weekday trips, and 50–60% of nonworking day trips (MLIT 2016), so one- or two-seater micro-EVs appeared suitable for most local trips. Such micro-vehicles are likely to reduce traffic congestion and parking problems, while also reducing the impact of accidents with pedestrians, due to their lower speed and weight. Moreover, their reduced energy consumption might make it easier to charge them with renewable electricity. According to MLIT, micro-mobility is well adapted for several types of users, such as the elderly or tourists (Lecler 2017). In 2012, ¥380 million (€3.1 million) was devoted for experiments jointly developed by localities and private companies. This was renewed up to ¥200 million (€1.6 million)/year (Lecler 2017). Among the 42 projects funded by the MLIT between 2013 and 2016, nine were one-way carsharing experiments, although most were very small. Other subsidies might have been obtained from other ministries or agencies by companies in different activity sectors. One example was a case in Kobe, which launched a service called Sea:Mo.[6] However, to our knowledge, experiments did not last after subsidies were terminated, with the exceptions of Choi Mobi and Ha:mo, which were both launched by car makers.

The Ha:mo Ride Experiment

TMC developed the concept of Harmonious Mobility Network (Ha:mo), including a multimodal information system (Ha:mo Navi) and a one-way carsharing service (Ha:mo Ride) aiming in addition to energy issue, to solve the last-mile problem from railways stations. Launched in 2012 in Toyota city with 10 vehicles (Coms), the service had 100 vehicles at 22 stations by October 2013 when it became a paid service. It also offered the possibility to use some i-Roads (only for round-trips) or rent electric bikes (100 bikes). The number of users increased from approximately 500 to 3710 in March 2015 with 35 stations. During the weekdays, the carsharing service's main purpose was commuting (45%), followed by professional trips (25%). The reasons for using Ha:mo were ease of use (64%), proximity to stations (60%), and saving time (33%). Walking and cycling (40%) were the main modes of transportation before using the one-way system, followed by private car (32%) and motorbike (13%) (Lecler 2017).

In a relatively small city (420,000 inhabitants), with low density in peripheral areas, and most people working in the car industry, the attractiveness of a carsharing system does not have the benefit of ideal conditions for its development. However, TMC led the firms' consortium of the smart community demonstrator and saw an opportunity to showcase the concept of urban micro-vehicles, promote Toyota Autobody Coms and its own i-Road, and test the business model of one-way carsharing. TMC does not consider carsharing as part of its business, but the carmaker did not terminate the service when the experiments ended in April 2015, even though a business model did not exist since operating costs were too high and the number of customers was too small.

Rather the service was extended and enlarged to 51 stations (21 without chargers). In addition, new experimental projects were launched in other locations, such as Times Car Plus × Ha:mo in Tokyo, which is the first Japanese experiment with one on-street station and the first one-way experiment for Times 24.

[6]This experiment was done in cooperation with Mitsubishi Heavy Industries (MHI), Nihon Unisys, Unitech, and Rokko Sangyo.

4.5 Analysis: Carsharing as a Public or a Commercial Service?

As these case studies show, the carsharing expansion in both countries followed quite different paths. Considered a car-rental service in both France and Japan, legislation required amending to allow for carsharing's commercial development, apart from experiments benefitting from special exemptions. How laws were modified, according to each country's vision of transportation, can be regarded as a path dependency and greatly influenced how carsharing evolved between round-trip and one-way systems with different uses and customers that corresponded to two distinct markets.

4.5.1 Round-Trip Versus One-Way

In the **round-trip-based system**, the attractiveness of the service depends on the proximity of stations to home and work locations, and implies 24/7 operation and enough fleet diversity to satisfy a variety of needs (small car for urban trips, family car for leisure, van for goods delivery). Round-trip is not designed for daily home-to-work trips, as the user would also pay for the parking time. The fully automated carsharing process for registering and picking up cars offered operators the ability to dramatically reduce labor costs compared to car rental. This was achieved with ICT development over the last 20 years. On average, the frequency of travel is low (one or two trips per month), for distances of 50–60 km and a rental period of 3 or 4 h. As trips are often in the late afternoon or on weekends (Lecler and Faivre d'Arcier 2014), operators tend, perhaps to a greater extent in Japan than in France, to prospect the companies whose trips take place during the day.

Round-trip is well adapted for people who do not own cars and can satisfy their daily mobility needs by alternative modes of transportation (PT, walking, cycling), that is to say living in dense urban areas. The high population density in Japanese cities, the correlated huge development of PT, and the prohibition of on-street parking are among the reasons why carsharing services meet a high demand in Japan. Conversely, high densities and well-developed PT alternatives in France are limited to city centers. On-street parking is possible and almost unrestricted in peripheral areas. Therefore, upscaling carsharing services, which is a condition for profitability, remains difficult as the LPA case showed. This explains why round-trip, which relies on conventional cars in both countries, is expanding faster in Japan than in France.

In a **one-way carsharing system**, members really appreciate paying for the service only when they are driving. According to public-policy objectives, one-way carsharing systems appear to be a new, complementary mobility service widening possibilities for inhabitants. However, some surveys show (6t-bureau-de-recherche 2014) that Autolib competes with PT, rather than strongly reduces private car ownership (Lecler and Faivre d'Arcier 2014). One of the main advantages of using carsharing is to avoid wasting time looking for parking, as a space is booked at the destination. However, one-way is much more costly and complex for operators to manage than round-trip, and even more so with EVs. Unlike round-trip systems, the frequency of use for one-way trips is higher (one or two trips per week, per user on average), but the duration of the rental is short (30 min) and the driving distance is limited (less than 10 km). Therefore, even more than for round-trip, conditions for one-way to be attractive must first include a large network of stations close to demand, in terms of origin-destination. Second, there must be guaranteed car availability at any time for all users. This means there must be a well-managed balance between empty and full stations, leading to additional labor costs to move cars from one station to another. These costs cannot be passed on to customers. Therefore, the one-way business model should be based on a greater number of rentals per vehicle, per day. More so than with round-trip carsharing systems, one-way systems are restricted to dense urban areas in which the size of the potential market offers hope for more users.

However, the different types of actors involved in each country also presumably played a part. In Japan, the main round-trip actors are companies engaged in parking and car-rental activities. Because these actors have parking spaces at their disposal that can easily be reserved for carsharing, implementation is less costly, especially if many of these spaces are at street level. Having experience managing car-rental businesses also helps these actors develop carsharing, especially round-trip, which is a rather similar business. If, as Japanese companies predict, car rental and carsharing will merge to offer a single membership service, even though fares differ between long- and short-term rentals, it may seem surprising that French car-rental companies do not engage more in carsharing. Times 24 recently developed Times Club, in which members can, with a unique card, choose between car rental and carsharing, depending on if the duration is longer than 10 h. The vision of carsharing as individual PT or commercial business might be part of the explanation, but the corporate structure is also at stake. Car rental and parking operations are not integrated in France as they are in Japan. Therefore, French car-rental companies, penalized by their limited office locations, seem as considering carsharing less attractive since margins are lower for short-term rentals. Public and private parking-management companies, with some exceptions (as shown by LPA Citiz Lyon), are also not involved in carsharing services, since they lack experience in car rental.

One-way is not following the trend of dynamic, private development of round-trip in Japan. Publicly subsidized experiments of one-way e-carsharing were performed, but they were so small that no positive network effects were found. These experiments often lasted no more than a year and couldn't attract many regular or active members. Apart from a willingness to test the system or the vehicles, most potential users were not willing to change their mobility behavior just for several months. As a result, these experiments all came to the same conclusion: There is no business model for one-way carsharing. Carmakers interested in promoting EVs, and ICT companies needing to gather data and test their models, engage in experiments for a short time. However, companies such as Times 24 or Orix, which are willing to develop carsharing as a commercial business, are not really enthusiastic. Although recognizing that clients might be interested in a one-way system, neither company intends, in a near future, to offer such systems. The MLIT sprinkling of funds to serve the largest number of localities ultimately does not allow for attracting enough practitioners to support the development of the services.

Even if station-based carsharing is expanding more in Japan than in France, the impact in terms of EVs diffusion, promoted by both governments and welcomed by cities to reduce CO_2 emissions and gain an environmentally friendly image, seems to work better in France. So, Autolib or Bluely in Lyon contributes to EV visibility and promotion. These services also give the opportunity to create a large network of charging stations for private EV owners, as one parking space at each station is reserved for such purpose. Although it is difficult to know for certain, as one-way in France is only proposed using EVs and only with micro-e-vehicles in Japan, it seems that EVs are better accepted by one-way users. One-way rental durations and driving distances are shorter, but at least a four-seater EV's range could also suit round-trip needs (average distance of 60–70 km). The conditions for an EV market to take off (battery capacity, charging infrastructure, and charging time) seem to similarly apply to carsharing for round-trip. For example, Times 24 had integrated some EVs into its round-trip fleet for a while. These EVs were replaced by conventional or hybrid cars for the following reasons: there were very few rentals, as customers did not want EVs; the purchasing or leasing cost of EVs is still much higher; EVs cannot be easily sold (or are too depreciated) on the underdeveloped second-hand EV market; and recharging needs also increased the investment burden, even though Times 24 installed charging devices in some of its car parks. Orix made some attempts as well, but finally also stopped for the same reasons. These examples confirm that for the diffusion of e-carsharing, cost is clearly an issue, but demand also matters.

French operators of round-trip, which is still not profitable, also do not seem to be considering introducing EVs into their fleet. Investing in purchasing (or leasing) higher-cost vehicles and charging devices, whether on-street or inside car parks, would further increase the financial burden. Therefore, holding and developing one-way e-carsharing services would require a partnership with cities and financial support, for example through public-service delegation contracts, as is the case for Autolib in the Paris region.

4.5.2 Two Visions of Transportation Between Public and Commercial Services

In France, collective transportation is considered a public service, so allocating public funds to guarantee a minimum of accessibility to all is accepted by populations and provided for by law. Since carsharing is defined as an individual public transportation, municipalities can design the service they need in relation to other locally available alternatives. The private operator must fulfill all requirements under the public-service delegation contract, but can limit financial risks either because part of the investment is covered by public funds or running costs are shared as long as the service is not profitable. Two French cities, Belfort and Pau, decided to manage a multimodal supply offering access to a PT network, a bike-sharing system, and round-trip carsharing. Belfort did so through its in-house operator, while Pau decided to operate its system through a public-service delegation contract (won by Keolis, a major PT operator). This can be related to the recent creation of the PMA, in charge of managing all mobility services in the urban area. Such a fully integrated system is justified as a means to offer a package of alternative services to car use and encourage reduced car ownership. Nevertheless, the risk of this public management lies in the vision that a public service should be available to all, leading to more stations in areas with low demand, resulting in low performance, and then in difficulties finding a business model.

In Japan, transportation is perceived as a commercial business that private companies operate for profit.[7] Using taxes to subsidize these businesses is not publically accepted, so is not provided for by law. This particularly applies to carsharing, which remains a car-rental service, even if regulations have relaxed to some extent. Therefore, public action to support new systems takes the form of experiments under government policy programs. However, it is difficult for cities, if not impossible, to financially support operators engaging in the business, or take over from national subsidies at the end of the experiment so that the service can be maintained on behalf of its social utility.

Therefore, changes do not address the same issues. In France, they aim to facilitate partnerships between cities and private operators, and in Japan to ease conditions for new services to be launched by private companies. In France, cities define suitable new mobility packages, which are operated under contract by private operators. In Japan, private companies engage in such packages by themselves, or through inter-companies partnerships, provided that a new demand or an expected future demand is high enough to ultimately be profitable. Developing carsharing in condominiums illustrates this enlarged vision of urban mobility. For example, Japanese real estate companies are soliciting carsharing companies to manage one to two cars as a means to compensate for the lack of parking spaces. It shows the interest of an integrated vision between transportation and city planning, but also environmental and energy issues. Therefore, although a demand exists, one-way carsharing, which does not appear able to become profitable within an acceptable time frame for private companies, has not yet taken root in Japan. However, the Japanese market-oriented approach avoids situation like that of Autolib in France, in which financial difficulties are partly due to underperforming stations that were

[7]It would be beyond the scope of this paper to explain this difference, rooted in each country socio-economic history.

requested by public authorities. It also has the advantage to overcome the problem of silo organization of public authorities, which penalizes new mobility services such as carsharing. Motivations for carsharing that are justified by environmental issues come from cities' environment departments, while PT is managed by transportation departments. Developing a new framework for the governance of smart, sustainable mobility (rather than transportation) implies breaking traditional frontiers based on technical competences within city departments. The perspective of Mobility-as-a-Service should be the opportunity to also think about reorganizing mobility management.

4.6 Conclusions

Carsharing appears to be a smart transportation system because it consumes less resource and copes better with individualized mobility needs. However, for it to develop as such, carsharing needs alternative modes of transportation (especially for home-to-work trips) performing well enough to encourage households to reduce car dependency. This is a condition for getting carsharing out of its current niche market, as it presently only satisfies a small percentage of mobility needs.

Let's come back to our research question: *Should station-based carsharing develop as a commercial or a public service, and what are the conditions required for it to take off?* It appears that the answer might depend on what is expected from the re-crafting or substitution to, the car driving practice. If reducing car ownership or use in densely populated cities is the only issue, then commercial services are probably the best means to recruit numerous "practitioners," as in Japan. However, it will be limited to round-trip, as one-way is too costly for companies to find a suitable business model. However, if one of the issues for carsharing is to promote environmentally friendly technologies, such as EVs, focusing on the last mile is necessary, which requires the use of one-way. If so, making carsharing a public service, allowing financial support from the public authority, is not only the best but in fact the only solution, as the French case has shown.

Moreover, even if station-based carsharing benefits from a certain defection from private car driving practice, it remains spatially limited. Indeed, whichever system is implemented or vehicles are offered, private initiatives will focus on areas with the highest potential market and best chances for profitability. This means low-density areas will not be attractive, while they are the source of high household car dependency. From a public-policy perspective, this means that reducing car use in France, or solving the aged population's mobility problem in Japan, will require public authorities to get involved in also defining a clear, financially bearable strategy for peripheral areas. It will probably take another form than market-mediated, station-based carsharing, unless autonomous car fleets develop fast enough to solve the problem of station proximity and unbalance.

References

6t-bureau-de-recherche (2014) Enquête sur l'autopartage en trace directe—L'autopartage en trace directe: Quelle alternative à la voiture particulière? Rapport final. ADEME, Paris

6t-bureau-de-recherche (2016) Enquête Nationale Autopartage—Mise à jour 2016—Analyse des enquêtes. ADEME, Paris

6t-bureau-de-recherche (2017) Enquête Nationale sur l'Autopartage—Mise à jour 2016—Etat des lieux technique et méthodologique. ADEME, Paris

APUR (2008) L'autopartage et autres modes alternatifs à la possession de la voiture particulière. http://www.apur.org/sites/default/files/documents/249.pdf. Accessed 22 Jan 2014

Autolib Métropole (2017) Autolib' Métropole avance avec vous, rapport d'activités 2016. https://drive.google.com/file/d/0B8MFxB5YvOOkUTBKM2ZMMy1yWVE/view. Accessed 12 Jan 2018

Bardhi F, Eckhardt GM (2012) Accessed-based consumption: the case of carsharing. J Consum Res 39(4):881–898. https://doi.org/10.1086/666376

Boston Consulting Group (2016) What's ahead for car sharing? https://www.bcg.com/fr-fr/publications/2016/automotive-whats-ahead-carsharing-new-mobility-its-impact-vehicle-sales.aspx. Accessed 21 Mar 2018

Cabinet Office (2003) Special zones for structural reform. Office for the promotion of special zones for structure reform. http://japan.kantei.go.jp/policy/kouzou2/sanko/030326setumei_e.pdf. Accessed 27 Jan 2018

CERTU (2008) L'autopartage en France et en Europe: état des lieux et perspectives. Lyon

Cervero R, Golub A, Nee B (2007) San Francisco City CarShare: longer-term travel-demand and car ownership impacts. Department of Transportation & Parking, City of San Francisco

Communauto (n.d.) Histoire de l'autopartage. http://www.communauto.com/historique01.html. Accessed 12 Feb 2018

Eco-Mo Foundation (2017) Zenkoku no kashearingu jirei ichiran (Overview of countrywide cases of carsharing). http://www.ecomo.or.jp/environment/carshare/data/carshare_jirei_2017.05.08.pdf. Accessed 5 Feb 2018

Faivre d'Arcier B, Lecler Y (2014) Promoting next generation vehicles in Japan: the smart communities and their experimentations. Int J Automot Technol Manage 14(3–4):324–346

Friedman D (1972) Public vehicle rental system: determination of feasibility. In: Selected proceedings of a conference on methods and concepts of forecasting travel demand for future systems. Transportation Studies Center, University of Pennsylvania, Philadelphia, pp 49–74

Geels FW (2012) A socio-technical analysis of low-carbon transitions: introducing the multi-level perspective into transport studies. J Transp Geogr 24:471–482. https://doi.org/10.1016/j.jtrangeo.2012.01.021

Kato H, Suzuki S, Kayama S et al (2015) Stakeholders' perspectives on the feasibility of their cooperation in the carsharing market: evidence from Japan. Asian Transp Stud 3(4):416–429

Lecler Y (2017) The promotion of micro-vehicle e-mobility in Japan. Paper presented at the 25th GERPISA international colloquium, R/Evolutions. New technologies and services in the automotive industry, ENS Cachan, Paris, 14–16 June 2017

Lecler Y, Faivre d'Arcier B (2014) Carsharing in cities: Will electric vehicles change the business? A comparison between France and Japan. Paper presented at the 22nd international colloquium of GERPISA, Kyoto, 4–6 June 2014. https://halshs.archives-ouvertes.fr/halshs-01092347/document. Accessed 12 Feb 2018

Loose W (2010) The state of European car sharing. MOMO Carsharing. Final report D 2.4 WP2. http://www.eltis.org/sites/default/files/tool/the_state_of_carsharing_europe.pdf. Accessed 22 Jan 2018

LPA Magazine (2017) Mobilités partagées: les atouts Lyonnais. http://www.lpa.fr/wp-content/uploads/2017/11/MOBART-n7.pdf. Accessed 22 Jan 2018

Martin E, Shaheen S (2011) The impact of carsharing on public transit and non-motorized travel: an exploration of North American carsharing survey data. Energies 4:2094–2114

Mizokami S, Nakamura K, Hashimoto J (2015) wanueigata MEV shearingu shisutemu no donyukanosei nikansuru shimureshon bunseki (Simulation analysis of the possibility to intro-duce one-way MEV sharing systems). J Jpn Soc Civ Eng 71(5):805–816. http://www.civil.kumamoto-u.ac.jp/keikaku/research/mizokami-pdf/file085.pdf#zoom=100. Accessed 20 Mar 2018

MLIT (2016) Choko mobiritei no seika to kongo (Micro-mobility, outcomes and hereafter). http://www.mlit.go.jp/common/001125685.pdf. Accessed 11 June 2016

Nakamura K, Mizokami S, Hashimoto J (2017) wanueigata kashearingu shisutemu no donyukanosei to saiteki suteshon haichi (The possibility of introducing one-way carsharing and the optimum location of stations). J Jpn Soc Civ Eng 73(3):135–147. http://www.civil.kumamoto-u.ac.jp/keikaku/research/mizokami-pdf/file089.pdf#zoom=100. Accessed 20 Mar 2018

Shaheen S (1999) Carlink—a smart carsharing system. J World Transp Policy Pract 5(3):121–128

Shaheen S, Chan ND (2015) Evolution of e-mobility in carsharing business models. In: Beeton D, Meyer G (eds) Electric vehicle business models. Springer, Berlin, pp 169–178

Shaheen S, Cohen A (2007) Growth in worldwide carsharing: an international comparison. Transp Res Rec 1992:81–89

Shaheen S, Cohen A (2013) Carsharing and personal vehicle services: worldwide market developments and emerging trends. Int J Sustain Transp 7:5–34

Shaheen S, Cohen A (2016) Regional and global carsharing market trends. In: Innovative mobility carsharing outlook. TSRC. University of California Berkeley. http://innovativemobility.org/wp-content/uploads/2016/02/Innovative-Mobility-Industry-Outlook_World-2016-Final.pdf. Accessed 21 Mar 2018

Spurling NJ, McMeekin A, Southerton D et al (2013) Interventions in practice: re-framing policy approaches to consumer behaviour. Sustainable practice research group report. http://www.sprg.ac.uk/uploads/sprg-report-sept-2013.pdf. Accessed 26 Mar 2018

Steininger K, Vogl C, Zettl R (1996) Carsharing organisations: the size of the market segment and revealed change in mobility behaviour. Transp Policy 3(4):177–185

Watson M (2012) How theories of practice can inform transition to a decarbonised transport system. J Transp Geogr 24:488–496

List of Interviews

Yokohama City: 2014/6/14, 2015/1/13
Nissan Motor (Choi Mobi): 2014/6/10, 2015/1/13
TMC (Ha:mo): 2013/3/13, 2015/1/21
Bluely: 2014/2/11, 2015/6/16
Lyon Métropole (urban mobility service): 2013/4/15, 2015/10/2, 2017/10/3
LPA Citiz: 2014/2/2, 2017/10/6
Times 24 (TCP): 2014/6/13, 2017/12/13
Rokko Sangyo (Sea:mo): 2017/11/29
MHI (Sea:mo): 2017/10/27
Orix Auto Corporation: 2017/12/11
Kobe City: 2017/11/10
Toyota City: 2013/3/13

MLIT (environment policy division and transport bureau): 2015/1/19
Osaka City (transport department and city planning department): 2017/11/13

Bruno Faivre d'Arcier is an emeritus professor in City Planning and Transport Studies at the Faculty of Economics and Management of the University of Lyon (Lyon 2) and is a researcher at the Transport Urban Planning Economics Laboratory (LAET) of the University of Lyon, France. Having been trained as an engineer and an economist, he started his career at the National Institute of Transport Research (Ministry of Transport), before joining the University in 1999, where he was the co-director of a Master's degree diploma on Urban and Regional Passenger Transport Studies. His research activity focuses mainly on local transport and mobility policies covering several approaches: Assessment of public policies (sustainable urban mobility plans) and transportation investments (cost-benefit analysis; public–private partnership); improvement of PT funding; analysis of PT service performance; and analysis of new mobility practices, such as electro-mobility, bike-sharing, and carsharing.

Yveline Lecler is an emeritus professor at Sciences-Po Lyon (University of Lyon) and senior research fellow at the Institute of East Asian Studies (CNRS, ENSL). She holds a Ph.D. in Social Sciences from EHESS (Paris) and a MA in Japanese language and civilization from INALCO (Paris). Specializing in Japanese political economics, her recent researches focus on innovation policies, in relation to energy management, low-carbon cities, next-generation vehicles, and e-mobility. She has written and edited several books, including *The Dynamics of Regional Innovation, the Policy Challenges in Europe and Japan*, co-edited with Yoshimoto Tetsuo and Fujimoto Takahiro (2012, World Scientific). She has also published many journal articles, including "Promoting next-generation vehicles in Japan: The smart communities and their experimentations" (with B. Faivre d'Arcier), in the *International Journal of Automotive Technology and Management*. She has been a visiting professor at several Japanese universities.

Part II
Governance of Automated Mobility

Chapter 5
Data-Led Governance of Self-driving Vehicles for Urban Shared Mobility

Tom Voege

Abstract The combination of big data, data analytics, ubiquitous mobile Internet access, and the business models enabled by these trends have led to a revolution in providing transportation and mobility services, particularly in dense urban areas. These services include various shared-mobility concepts and demand-matching platforms for e-hailing and other Transportation Network Companies (TNCs). While these have been highly successful in many global markets, there are also many well-documented and cases in other regional markets in which noncompliance with regulatory frameworks governing legacy transportation modes led to bans and protests. Add in vehicle automation, which appears to be the ultimate aim in the mobility service industry, due to large cost-savings potential, and current regulatory approaches will likely reach a breaking point. Data-driven governance is increasingly seen as a way forward, giving more flexible regulatory environments catering to the rapid innovation we are witnessing in many sectors. This can be a solution for the current regulatory deadlock in transportation policy and emerging shared-mobility services, allowing systems contributing to the public good to be rolled out in a safe, flexible setting, enabling market uptake while protecting public safety and integrating legacy providers.

Keywords Automated vehicles · Shared mobility · Regulatory frameworks
Disruptive innovation · Data-led governance

5.1 Introduction

5.1.1 Background Discussion

We are now at a stage in which implementation of self-driving vehicles (SDVs) is a clear trend. While projected timescales, technology options, and use cases vary, it is

T. Voege (✉)
ITF-OECD, Paris, France
e-mail: tom.voege@itf-oecd.org

M. Finger and M. Audouin (eds.), *The Governance of Smart Transportation Systems*, The Urban Book Series, https://doi.org/10.1007/978-3-319-96526-0_5

necessary for policy-makers to prepare their response. Many benefits of vehicle automation have been suggested, including improved road safety, decreased emissions, and increased network capacity. The emergence of related shared-mobility services enabled by SDVs also promises even larger benefits for society as a whole and at the city level.

However, negative effects could also be envisaged, including network-capacity gains, leading to induced traffic; the ability of more productive travel time, leading to longer trips with people moving further away from commercial centers; and potentially huge fleets of empty vehicles running errands and generating much more congestion. Many governments are heavily investing in research and development (R&D), and demonstration of near market-ready systems, showcasing their ambitions for leadership in this space. In parallel, younger, emerging companies, with a strong information technology (IT) focus and leadership mentality, are aggressively pushing related systems and services into the market.

Therefore, automated vehicles (AVs) are increasingly becoming part of the sharing economy and disrupting technologies, in both the negative and the positive sense. Policy-makers' key role will actively manage the transition period, which has already begun. Legal and regulatory frameworks are key tools for this. The current industry and academic discourse on SDVs usually identify regulation as a key barrier to wider implementation of an otherwise mature technology. This often technology-led discussion might be overly optimistic, but it is a realistic assumption that the technology is mature for a number of environments and uses. Therefore, leadership from policy-makers is both timely and essential.

The discussion about regulatory frameworks in the context of SDVs typically centers on the vehicles, or the vehicle-type approval processes for guaranteeing road safety. Work is ongoing on many national and international levels, including concepts for SDV driving tests, elaborate test tracks simulating various real-life scenarios, and governments amending legal frameworks to allow testing on public roads. While regulating the automotive aspect of SDVs is vital, given the likely implementation of this technology to enable shared-mobility concepts, regulating mobility services must be considered in parallel. Such services will directly compete with legacy transportation services, which are often heavily regulated and protected.

Highly publicized issues in terms of regulating technologies and services are very visible in the case of transportation network companies (TNCs), but also deal with the functionalities of privately owned SDVs on public roads. Such issues are often due to an increasing time gap between innovation and related regulatory responses. Policy-makers are under increasing pressure to find a balance between administrating oversights and enabling innovation. The advent of big data and data science (particularly the blockchain approach), and its application to the transportation sector, has the potential to solve this dilemma by creating much more flexible, data-led regulatory approaches to ensure specific policy objectives are followed.

In multi-modal public-transportation journeys, the last mile is crucial. Lack of convenience and safety concerns for this trip segment often deter modal shifts.

Conventional public transportation is unable to provide last-mile transportation in many cases, particularly at low-demand times and in low-density locations. Shared-mobility concepts and vehicle automation particularly have the potential to radically improve service provision, enabling a paradigm shift for urban mobility. The potential for mobility solutions, such as carsharing, carpooling and e-hailing to meet urban transportation demand, is attracting increasing attention. We might be seeing the first waves of a radical change in formats for car use and ownership, and overall mobility provision in urban areas.

5.1.2 Related Key Trends

The motivation for the discussion on data-led governance of SDVs for urban-shared mobility is that mobility in general is currently at a crossroads of being affected by various trends, including the sharing economy and disruptive innovation, which in the case of mobility platforms are different sides of the same coin. The sharing economy trend relates to shared mobility concepts, including new business models allowing use and shared access to vehicles. The disruptive innovation trend relates to e-hailing and other services commonly referred to as TNCs, and the rise of platform-based mobility services.

The *Oxford Dictionary* defines the term sharing economy as "*an economic system in which assets or services are shared between private individuals, either free or for a fee, typically by means of the Internet*". The key driving forces behind the rise of sharing economy include information technology and social media, social and online commerce, and increasing volatility in cost of natural resources. In terms of the services and uses offered to end users, the sharing economy comes in many different shapes and forms. For example, Botsman and Rogers (2011) have defined the following:

- *Product–service systems*: allow consumers to engage in monetized exchanges through peer-to-peer temporary access to goods. Privately owned goods can be shared or rented via peer-to-peer marketplaces. For example, DriveNow is a car-rental service that offers an alternative to owning a car through usage-based pricing.
- *Redistribution markets*: are systems of collaborative consumption based on used or pre-owned goods passed on to someone who wants them. This is another alternative to the more common "reduce, reuse, recycle, repair" methods of dealing with waste. In some markets, the goods may be free. In other markets, the goods are swapped or sold for cash.
- *Collaborative lifestyles*: allow consumers to engage in monetized exchanges through peer-to-peer services or access to resources such as money or skills. These systems are based on people with similar needs or interests banding together to mutualize and exchange less-tangible assets such as time, space, skills, and money. The growth of mobile technology provides a platform to enable location-based GPS technology and provides real-time sharing.

While this definition and its variations have been widely used and accepted, it is not without controversy. Sundararajan (2013), among others, have argued that the term *sharing economy* is a misnomer. *Harvard Business Review* suggested the correct term, in the broad sense, should be *access economy*. The authors argued that when sharing is market-mediated, and a company is an intermediary between consumers who do not know each other, it is no longer sharing. Instead, consumers are paying to access someone else's goods or services.

Closely related to product–service systems and the underlying commercial, peer-to-peer mutualization systems (CPMS), shared mobility (Shaheen et al. 2016) has emerged through the principles of the sharing economy being applied to the passenger transportation and mobility sector. As with other aspects of the sharing economy, shared mobility also is controversial.

The disruptive element of shared mobility is seen in the shape of transportation network companies (TNCs), particularly Uber, both in terms of its position vis-à-vis legacy transportation providers (particularly taxis) and its conflict with regulatory frameworks (which are often both protectionist and aim to safeguard the wider public goods that unchecked markets do not support). This will be the subject of the following section. This section will discuss shared access to, and use of vehicles, which decision-makers, media, and civil society commonly frame in a positive manner.

On the other hand, *disruptive innovation* is a business administration term that refers to an innovation that creates a new market and value network, eventually disrupts an existing market and value network, and displaces established market-leading firms, products, and alliances. The term was defined and first analyzed by Christensen (1997). Disruptive innovations tend to be produced by outsiders and entrepreneurs, rather than existing market-leading companies.

The business environment of market leaders does not allow them to pursue disruptive innovations at first because they are not profitable enough and their development can take scarce resources away from sustaining innovations that are necessary to compete against current competition. A disruptive process can take longer to develop than by the conventional approach, and the associated risk is higher than with other, more incremental or evolutionary forms of innovations, but once it is deployed in the market, it achieves a much faster penetration and higher degree of impact on the established markets.

The previous section discussed the less-controversial aspects of the emerging new mobility services, characterizing them as examples of the generally more positively framed sharing economy. On the other hand, e-hailing is a clear example of a more disruptive element, and is also the subject of this section. Examples include conflicts with existing regulatory frameworks, perceptions of unfair competition with legacy transportation services, concerns over road safety and passenger personal safety, and potentially negative labor market effects. These examples were all made very public through the media reporting on demonstrations and protests over the last years.

5.1.3 Aims and Methodology

Against the backdrop of these key trends, there must be a pragmatic discussion of current regulatory frameworks governing vehicle design and the provision of mobility services as a basis for recommending a way forward. The key research question asks if data-led governance of AVs for urban-shared mobility is a way forward, both in terms of finding a more flexible regulatory regime allowing quicker uptake of innovative solutions, but also solving the regulatory conflict between new players (such as Uber) and legacy operators (such as taxis)? The move of mobility platforms and other mobility service providers to the use of SDVs is intuitive, given the large cost-saving potential. The first large trials are already taking place, such as Waymo or the Uber/Volvo partnership.

The qualitative empirical basis for this analysis includes a combination of desk research, engagement of various external experts and stakeholders, and discussion at workshops and roundtables with internationally renowned experts in academia, R&D, regulators, government agencies, and industry. These were carried out by the International Transport Forum (ITF) of the Organisation for Economic Co-operation and Development (OECD), through projects from the program of work of both its Transport Research Centre (TRC) of Member Countries and its Corporate Partnership Board (CPB) of Industry Partners. Relevant policy analysis projects and publications included:

- ITF CPB Report Big Data and Transport—Understanding and Assessing Options (2015a)
- ITF CPB Report Data-Driven Transport Policy (2016a)
- ITF CPB Report App-Based Ride and Taxi Services: Principles for Regulation (2016b)
- ITF Roundtable on Co-operative Mobility Systems and Automated Driving (2018)
- ITF Case-Specific Policy Analysis Report Regulation of For-Hire Passenger Transport: Portugal in International Comparison (2016c)
- ITF CPB Report Data-led Governance of Road Freight Transport—Improving Compliance (2017)
- Ongoing work of the ITF Working Group on Big Data and Open Data
- Ongoing work of the OECD Going Digital Project

The analysis of the key research question posed above and the challenges and opportunities involved are based on discussions, in the following chapters, of new mobility concepts and sharing platforms, the impacts of vehicle automation technology, and the potential of big transportation data and data-led governance, followed by a concluding discussion giving an outlook of future research needs.

5.2 New Mobility Concepts and Sharing Platforms

5.2.1 Defining Systems and Concept

Shared mobility is still a relatively new field, so business models and preferred technologies are still in flux. According to a TCRP Research Report (TRB 2017), current systems and services include the following: bike sharing, car sharing, demand-responsive transportation systems, fixed-route systems, micro-transit, mobility on demand, para-transit, private shuttles, public transportation, ride-sharing, carpooling, ride-sourcing, ride-splitting, dynamic carpooling, and specified public transportation. Shared mobility was generally defined as a wide range of transportation services having in common that they are shared among users.

A study by management consultancy McKinsey on new business models and technologies for solving mobility challenges (2015) compared characteristics of mobility solutions (Table 5.1). The type of transportation services currently provided through the app-based platforms provided by TNCs, such as matching passengers with drivers, operate in the traditional for-hire passenger transportation services market. These include taxis, for-hire cars with drivers, and some forms of on-demand micro-transit. These types of transportation services have a long history pre-dating public transportation as a feature of large- and medium-sized cities around the world. They are an essential component of well-functioning mobility in metropolitan areas, as they deliver convenient, door-to-door trips for those who require such services either regularly or quickly. Although they only account for a small share of overall trips, they are always essential for some people, at some times, under some circumstances. For-hire transportation services are a stand-alone part of the urban mobility offerings but can also be an important addition to

Table 5.1 Characterization of mobility solutions

	Traditional mobility solutions	New mobility services	
Individual mobility	Private car ownership	Car sharing: peer-to-peer	A platform for individuals to rent out private vehicles when not in use
	Rental cars	Car sharing: fleet operator	On-demand, short-term car rentals with vehicles owned and managed by fleet operator
Group mobility	Public transportation: group mobility	On-demand private shuttle	App- and technology-enabled, cheaper than taxi, more convenient than public transportation
	Public transportation: mass transit	Private buses	Use of shared commuter-bus fleets available to the public or to company employees

Adapted from McKinsey (2015)

multi-modal public transportation provision, walking, cycling, individual cars, or carsharing services. However, they are now facing significant disruption, due to the arrival of new mobility services using app-based platforms. These services are known by several names, including ride-sourcing companies, ride-booking, on-demand IT-based transportation aggregators, commercial-transportation intermediaries, and transportation network companies.

TNCs connect via Web sites and mobile apps, pairing passengers with drivers who provide transportation in non-commercial vehicles. Examples of TNCs include Gett, Lyft, Juno, Cabify, Uber, goCatch, Via, Ola Cabs, GoCar, GO-JEK, Careem, Wingz, Taxify, GrabTaxi, Didi Chuxing, Easy Taxi, 99, and Fasten. The term TNC originated from the California Public Utilities Commission (CPUC 2013), which defined a transportation network company as "*a company that uses an online-enabled platform to connect passengers with drivers using their personal, non-commercial, vehicles*". The state of Virginia defines a TNC as a company that "*provides prearranged rides for compensation using a digital platform that connects passengers with drivers using a personal vehicle*".

TNC platforms have sometimes been called ride-sharing, but the terms e-hailing and ride-sourcing have been developed to describe the transportation services associated with TNCs. Some early reports used the latter term to clarify that drivers do not share a destination with their passengers, and that the driver's primary motivation was income. The term ride-sourcing means outsourcing rides. However, we prefer e-hailing to talk about Lyft- and Uber-like services, as it illustrates the digital nature of such solutions, and the availability of private vehicles and taxi services on the platforms.

5.2.2 Emerging Regulatory Conflicts

The popularity of TNCs is becoming increasingly evident, not only by the speed of consumer uptake in cities around the world, but also by how TNCs have mobilized significant sums of venture capital backing and formed global alliances enabling new configuration of urban transportation services. There is considerable interest in how these services may affect urban transportation markets, including public transportation, and might ultimately reshape the cities in which they operate. Because of these potential impacts and the fact that the arrival of TNCs has disrupted highly regulated incumbents in most cities, public authorities have been under intense pressure to regulate these new services. The pace of change has been rapid, with the first TNC operations starting in 2009 in the US, and soon spreading around the world. Early innovators in the field were joined by numerous competitors that have sought to leverage the same technologies and similar business models. In response, incumbents have deployed apps to retain and increase patronage. In many cases, the rapid pace of change has caught authorities off-guard, as TNCs typically do not fall under existing regulatory structures, and the deployment of new technologies blurs the line between existing regulatory categories. Some TNCs have taken advantage of this regulatory ambiguity to deploy services that challenge the spirit of existing rules in the minds of many regulators. In response, regulators have typically sought either to block market entry or adapt existing regulations with more-or-less success.

However, public authorities often fail to address the fact that TNCs can deliver real efficiency gains and consumer benefits, and are attractive to many workers, including those seeking self-directed and flexible working arrangements. Therefore, TNCs are a growing, permanent feature in many cities around the world. Few countries have looked at completely overhauling existing regulations to adapt the regulatory environment governing for-hire services to new market conditions, but there are indications that some may soon do so. Another shortcoming is that such incremental responses fail to anticipate and provide a sufficiently flexible regulatory framework for even greater disruptions building on the horizon. Foremost among these is the likely combination of TNC-like e-hailing services with highly automated and, ultimately fully automated, fleets of vehicles.

However, the likely use of SDVs for shared-mobility systems necessitates parallel, simultaneous consideration of public transportation regulations. This will be discussed in the following section.

5.2.3 General Regulatory Principles

Regulating public transportation must strike a balance between encouraging entrepreneurship of operators, ensuring the social dimension of public transportation, and ensuring service quality and customer satisfaction. Levels of regulation (and active de-regulation) and policy support for public transportation varies widely around the world. Table 5.2 shows an overview of general public transportation-policy stakeholders and responsibilities, according to UITP training material. The relevant general tasks of a transportation authority, in the context of regulatory frameworks and wider transportation policy considerations for public transportation services, include the following:

- *Market organization*: Elaborate market regulation, procurement strategy, and standard contract; give advice on the regulatory framework.
- *Management of budgets and assets*: Elaborate fare revenues and operating costs; define fares; define ownership of infrastructure and fleet systems.
- *Defining level of service*: Define the route network and transportation modes; define operating hours and frequencies per route.
- *Monitoring and payments*: Monitor operator performance; determine service fees, bonuses, and penalties.
- *Integrated approach*: Ticketing; information (online and offline); and marketing and promotional campaigns.
- *Infrastructure*: bus terminals/bus stops, dedicated bus lanes and priority, traffic management, parking management, non-motorized transportation.

Table 5.2 Public transportation-policy stakeholders and responsibilities (author's elaboration, based on UITP material)

Level	Actors	Role
Strategic (long term)	National government	•Legal framework •National transportation policy •Financial support local government
	Local government	•Transportation policy •Budget •Fare policy •Local regulations
Tactical (medium term)	Transportation authority	•Public transportation network •Service levels •Contracting operators •Ticketing system •Information and marketing •Investments in infrastructure
Operational (short term)	Transportation operators	•Transportation operation •Investments in rolling stock •Customer services

In the advent of emerging shared mobility concepts using SDVs, traditional regulatory frameworks, which are already stretched to accommodate current advances, might reach a breaking point, requiring novel approaches and systems.

5.3 The Impacts of Vehicle Automation Technology

5.3.1 History of SDV Technology

The concept of SDVs can be traced back to the 1939 World Fair in New York and the General Motors (GM) Futurama exhibit, which included the concept of an automated highway system. This has been a firm element of a vision of the future in popular culture since then. Given that this was the anticipated progress within the next 25 years, by 1964, actual developments in this field have somewhat lagged. Nevertheless, 1949 saw the first experiments with a remote-controlled car, as a joint effort between GM and RCA.

It was not until the 1980s that real progress with vehicle automation was made, enabled by breakthrough developments in computing, sensor, and communication technologies. In 1986, the California Partners for Advanced Transit and Highways (PATH) program began, which was a collaborative project between Caltrans and UC Berkeley. It carried out a wide range of transportation research, including one of the earliest vehicle automation projects. At the same time, the EU-funded PROMETHEUS Project (Program for European Traffic of the Highest Efficiency and Unprecedented Safety) developed and tested similar concepts and technologies (Shladover 2017).

Developments accelerated further in the new millennium with automation and platooning demonstrations in the USA and Europe. A series of EU-funded projects (CyberCars, CyberMove, NetMobil, and CityMobil) that started in 2001 specifically looked at automated, urban, shared mobility. The DARPA Challenges in the USA between 2004 and 2007 gave a further boost to R&D efforts, enhancing public visibility. However, it was not until Google's announcement in 2010 of its efforts to test and develop a self-driving car (now Waymo), that the automated car achieved its current hype status. Governments, cities, and companies around the world are now competing for leadership in this field.

Although a large number of systems and technologies are being tested and demonstrated, providing a wealth of data and information, it is still too early to tell which specific solutions will eventually emerge. Despite this uncertainty, it is essential for policy-makers to positively influence these developments in the transition period to automated transportation. Using SDV technology to provide shared-mobility concepts in urban areas is a key trend, based on ongoing research projects, demonstrations, and studies. By increasing average vehicle occupancy rates, automated shared mobility services promise to reduce the number of vehicles on the road and ultimately provide more sustainable mobility, which could eventually decrease congestion and free up parking spaces for more appropriate use (ITF 2015b), at least to the extent that the space would not be occupied by pent-up demand for more road traffic.

5.3.2 Current SDV Developments

Discussion of automating road vehicles requires establishing some definitions of terminology. While a variety of different technology options, application environments, and business models can be envisaged, a useful categorization can be made, based on the roles and responsibilities of the driver and/or passenger in the vehicle.

According to the updated SAE International (formerly US Society of Automotive Engineers) *Taxonomy and Definitions for Terms Related to Driving Automation Systems for On-Road Motor Vehicles* (SAE 2016), the varying levels of automation are to be classified as follows:

- *SAE Level 0 (No Automation)*: full-time performance by the human driver of all aspects of the dynamic driving task, even when enhanced by warning or intervention systems.
- *SAE Level 1 (Drive Assistance)*: driving mode-specific execution by a driver assistance system of either steering or acceleration/deceleration using information about the driving environment and with the expectation that the human driver perform all remaining aspects of the dynamic driving task.
- *SAE Level 2 (Partial Automation)*: driving mode-specific execution by one or more driver assistance systems of both steering and acceleration/deceleration

using information about the driving environment and with the expectation that the human driver perform all remaining aspects of the dynamic driving task.
- *SAE Level 3 (Conditional Automation)*: driving mode-specific performance by an automated driving system of all aspects of the dynamic driving task with the expectation that the human driver will respond appropriately to a request to intervene.
- *SAE Level 4 (High Automation)*: driving mode-specific performance by an automated driving system of all aspects of the dynamic driving task, even if a human driver does not respond appropriately to a request to intervene.
- *SAE Level 5 (Full Automation)*: full-time performance by an automated driving system of all aspects of the dynamic driving task under all roadway and environmental conditions that can be managed by a human driver.

As Bishop and Shladover (2015) note, road transportation automation systems are not ends in themselves, but are a means of satisfying needs to improve transportation operations or driver comfort and convenience. Specific systems will be designed to achieve different goals that are likely to point toward very different designs. These goals could include combinations of enhancing driving comfort and convenience; improving productivity or quality of life by freeing up time; reducing vehicle-user costs; improving vehicle-user safety or broader traffic safety; reducing travel time; enhancing/broadening mobility options; giving users more flexibility; reducing traffic congestion; reducing energy use and pollution emissions; making more efficient use of existing road infrastructure; and reducing the cost of future infrastructure and equipment.

Walker-Smith (2014) defined one simple means of understanding the opposing approaches to deploying SDVs:

- *Everything-somewhere* (such as Waymo): Very high automation in a constrained geographical area, due to the need to constantly update mapping and limit the interactions with potentially hazardous (higher-speed) traffic. Given the high functionality, it is likely that the fleet would need frequent servicing and testing to ensure safe operation is maintained. This is also facilitated by geographic constraints.
- *Something-everywhere* (such as automotive OEMs): This is the classic incremental approach, in which systems are brought to market that are capable of operating on "any" road (of a certain type, at least) regardless of geographic area.

Some OEMs advocate another approach, which could be termed a *something-eventually-everywhere* scenario. This includes sections of roadway individually approved for automated operation by the OEM and/or public authorities, based on availability of map information and potentially by modifications to the supporting infrastructure, as required by public safety agencies and/or automation system developers. This may entail the vehicle traveling the route first to collect map information to support the on-board system.

One of the first real-world applications of SDVs was the ParkShuttle in Rivium Business Park, near Rotterdam, in the Netherlands. This was a shared, automated shuttle carrying passengers on a loop, with a number of stops connecting offices with a nearby public transportation interchange. This system opened to the public in 1999. A similar system connecting the long-term car park with the main terminal building of Schiphol airport has been set up as a multi-year demonstration.

Both systems, although physically segregated from other traffic on some parts of the route, operated safely with manually driven vehicles and pedestrians in controlled, but mixed, environments. The Rivium system has been extended, with second-generation vehicles and is still operational today. Similar kinds of shuttles have since been tested around the world (such as in Paris, Las Vegas, and Singapore). There are currently about 10 different suppliers in the market. Most of these systems have similar parameters, including:

- *Low speeds*: Operational speeds are often below 15 km/h, but higher speeds are possible, depending on segregation, as maximum levels of jerk for emergency braking must be followed for safety reasons.
- *Simple, controlled environments*: Minimized interaction with other traffic, operation such as in pedestrian zones and on university campuses.
- *Significant infrastructure*: There are varying levels of physical separation of AVs from other traffic (such as manually operated, vulnerable users).
- *Supervised operations*: While it is not necessary to have operations staff inside the vehicle, a staffed remote operations center is necessary.

Notable first concepts of integrating the e-hailing aspects of TNCs (or at least providing a transportation service) with highly automated vehicles include the US cooperative efforts between Uber and Volvo, and Lyft and Ford; the demonstration projects of Navya/Easymile shuttles around the world; and the Transport Catapult Initiative testing carried out by the UK government. These examples show a mix of TNCs partnering with traditional OEMs, as well as the use of purpose-built vehicles by new players entering the market. For a move from early pilot programs to widespread applications, as in the case of standalone TNC systems, positioning of these services vis-à-vis existing regulatory frameworks must be revisited in the light of changed circumstances.

5.3.3 General Regulatory Principles

The current discussion about the applicable regulatory frameworks in the context of AVs typically centers on the vehicle itself, in other words, the type-approval processes for guaranteeing road safety. This work is ongoing at national and international (in particular, UNECE WP.29, see below) levels, including concepts for AV driving tests, elaborate test tracks simulating various real-life scenarios, and governments amending legal frameworks to allow testing on public roads.

While regulating the automotive aspect of SDVs is vital, given the likely implementation of this technology for shared-mobility concepts, the regulation of mobility services must be considered in parallel, as they will be in direct competition with legacy transportation services, which are often heavily regulated and protected, as discussed in the previous section.

The World Forum for Harmonization of Vehicle Regulations is a working party (WP.29) of the Sustainable Transport Division of the United Nations Economic Commission for Europe (UNECE). It is tasked with creating a uniform system of UN regulations, for vehicle design to facilitate international trade. WP.29 was established in June 1952 as the *Working Party of Experts on Technical Requirement of Vehicles*, while its current name was adopted in 2000. The forum works on regulations covering vehicle safety, environmental protection, energy efficiency and theft-resistance. Key legislations include the 1958 and 1998 agreements. The UNECE (1958) agreement operates on the principles of type approval and reciprocal recognition. Any country that accedes to it has the authority to test and approve any manufacturer's design of a regulated product, regardless of the country in which that component was produced. Each individual design from each manufacturer is counted as one individual type. Once any acceding country grants a type approval, every other acceding country must honor it and regard that vehicle or item as legal for import, sale, and use. The UNECE (1998) agreement aims to produce meta-regulations, called global technical regulations (UN GTRs), without administrative procedures for type approval; meaning without the principle of mutually recognizing type approvals. The 1998 agreement stipulates that contracting parties must establish UN GTRs in a UN global registry by consensus vote. The UN GTRs contain globally standardized performance requirements and test procedures.

UNECE WP.29's work is ongoing to find a regulatory response to SDVs. It currently follows the evolutionary scenario (rather than the disruptive one) favored by OEMs, in which higher levels of vehicle automation are part of developing advanced driver assistance systems (ADAS). Such systems include adaptive cruise control or parking-assistance systems, which can be subject to the UNECE regulatory system, as already practiced.

5.4 The Potential for Data-Led Governance

5.4.1 Defining Big Data

The term *big data* is commonly defined as high-volume, high-velocity, and/or high-variety information assets that demand cost-effective, innovative forms of information processing that enable enhanced insight, decision-making, and process automation. While the term is relatively new, the act of gathering and storing large amounts of information for eventual analysis is not. The concept gained momentum in the early 2000s when Laney (2001) articulated the now-mainstream definition of *big data* as the three Vs:

- *Volume*: Organizations collect data from a variety of sources, including business transactions, social media, and sensors or machine-to-machine data. Storing it would have been a problem in the past, but new technologies (such as Hadoop) have eased the burden.
- *Velocity*: Data streams in at an unprecedented speed and must be handled in a timely manner. RFID tags, sensors, and smart metering are driving the need to handle torrents of data in near real time.
- *Variety*: Data comes in all types of formats, from structured, numeric data in traditional databases, to unstructured text documents, email, video, audio, stock ticker data, and financial transactions.

Statistics software and IT company, SAS, considers two additional dimensions in the context of big data, acknowledging that the field of data science is still in flux:

- *Variability*: In addition to increasing data velocity and variety, data flows can be highly inconsistent with periodic peaks, such as trending topics on social media. Daily, seasonal, and event-triggered peak data loads can be challenging to manage; even more so with unstructured data.
- *Complexity*: Data currently comes from multiple sources, which makes it very difficult to link, match, cleanse, and transform it across systems. However, it is necessary to connect and correlate relationships, hierarchies and multiple linkages, or data can quickly spiral out of control.

5.4.2 Big Transportation Data

The growing importance of big data, beyond the initial hype, led many government agencies and private-sector companies to investigate the opportunities and challenges of a wider move toward the digital(-ized) economy. The ongoing digital transformation of economies and societies holds many promises to spur innovation, generate efficiencies, and improve services. In doing so, this transformation could boost inclusive and sustainable growth, as well as enhance well-being. However, these benefits go hand in hand with disruptions. The nature and structure of organizations and markets are being transformed, raising important issues about jobs and skills, privacy protection, cyber-security, and how to ensure that technological changes benefit society as a whole.

For example, the OECD Going Digital project is aimed at bringing about stronger, more inclusive growth from the digital revolution, for which it is essential to build a coherent, comprehensive policy approach. Moreover, some countries and groups are better placed than others to harness the benefits of digital transformation. When applying the concept of big data to the transportation and mobility sector, or big transportation data, specific challenges and opportunities arise, in addition to various common horizontal themes.

According to the UK Department for Transport (2017), the transportation sector has always collected and analyzed large quantities of data, such as that from timetables, traffic news, and air schedules. However, recent developments in the quantity, complexity, and availability of data collected from, and about, transportation, together with advances in computing technology, are presenting new opportunities to create more efficient, smarter transportation systems for people and freight. In addition, opening up data in the transportation and mobility sector by making it more widely available, and linking it with data from other sectors, is part of government strategies in many countries to improve transparency and encourage economic growth.

The growth of technology in the transportation sector has led to an increase in the type and amount of data automatically created and collected about moving people and freight. Examples include smart travel cards, sensor equipment embedded in the built environment, location tracking through global positioning systems (GPSs), and mobile networks. Using big data can increase efficiency and reduce costs to infrastructure and service operators, and provide better levels of services to users. Examples include key applications for maintaining assets, managing road traffic, planning public transportation services, and informing user decisions.

5.4.3 To Data-Led Governance

A key application of big data analytics going forward is its use for data-driven, data-led, evidence-based, regulatory frameworks. There is a particular opportunity to use big data for flexibly regulating transportation services that big data enables. Such an approach would necessitate translating current descriptive regulations into quantifiable targets or indicators. Regulator access to data (either mandated or through data-sharing partnerships) could enable a dash-board approach using data analytics, in which instances of crossing pre-defined policy or performance thresholds would be flagged and visualized.

The ITF report on *App-Based Ride and Taxi Services: Principles for Regulation* (2016b) discussed the specifics of data-led TNCs and TNC-like service regulation. Policy should allow development of innovative services toward public-policy objectives to equitably improve mobility, safety, consumer welfare, and sustainability. This principle will likely include lightening the regulatory controls on entry and fares for dispatched taxi services, while controlling for relevant public-policy objectives and introducing newer types of regulation that are currently deployed in most markets.

As much as possible, regulators and regulation should not differentiate among categories of service providers. If exemptions are required, these should be explicitly substantiated and regularly reviewed. To the extent that technology or other alternative approaches can deliver on public-policy objectives in place of traditional and currently required regulatory frameworks (including consumer protection), regulations should be adapted, streamlined, replaced, or removed, based on open, evidence-based analysis.

Regulators should increasingly seek to leverage data-led regulation as a way to provide a lighter and more flexible approach to achieving public-policy objectives. Operators and platforms should be able to choose lighter, more flexible regulatory approaches versus wider spectrum, more burdensome regulatory frameworks as long as they provide regulators with information that allows authorities to deliver public-policy outcomes.

This control can be based on access to data, reception of, or access to, vetted, third-party data-reporting systems that allow regulators to seek specific insight into how for-hire operations are affecting public-policy objectives, preventing market abuse or contributing to congestion, based on open, evidence-based analysis. Policy-makers should identify minimum data inputs from operators to carry out the above analyses and, in particular, plan and regulate mobility services in line with public-policy objectives.

5.4.4 Government Role in Regulating SDV

We have discussed the emergence of new data-enabled mobility systems, their positive and controversial aspects, specific regulatory challenges, and a potential way forward using data-led approaches for a new comprehensive regulatory framework. The next step is to see how these general ideas can be applied to the regulatory response to join the trends of SDVs and shared mobility into automated shared mobility.

While SDVs are undeniably a major trend, it is still too early to tell how much of it is purely hype (which might evaporate), which current implementation scenarios will become a reality, and which additional ones will emerge. Current development points toward competing pathways, which comprise an incremental approach some OEMs favor. This approach has much less transformational potential for current transportation provision. The disruptive approach would see a paradigm shift of transportation and mobility, replacing most (if not all) private car trips with shared-mobility services. A roundtable on the use of SDVs for urban-shared mobility carried out by the ITF (2018), discussed potential government actions in this context.

SDVs' potential to improve road safety and mobility, while reducing congestion, has generated substantial excitement. However, most experts agree that society's ability to capture these benefits and minimize negative impacts depends on the presence of effective regulatory frameworks. Current vehicle safety regulations can be adequately stretched (if imperfectly) to accommodate the early commercial deployment of SDVs. In turn, regulators will be able to use real-world evidence to ensure that SDVs are safely deployed.

Stretching existing regulatory frameworks will not be sufficient in the long term. Once SDVs become a critical proportion of all motor vehicles, regulatory frameworks that reflect the capabilities of SDVs and the nature of the industry will be necessary. In particular, road rules will need to change and compliance mechanisms related to them will need to shift from an *ex-post* criminal enforcement approach to an *ex-ante* product safety approach. In the meantime, the research and non-regulatory/quasi-regulatory approaches some governments adopted provide useful examples of good short-term practices.

While road safety is a key concern for SDV deployment, there is increasing evidence that regulators must start preparing for the broader impacts of SDVs on society as a whole. Many of the potential benefits from introducing SDVs depend on a large modal shift from the private car to shared-mobility services. In the absence of regulatory interventions, experts consider this scenario to be unlikely, due to dense urban areas where individual vehicle ownership may already be low and high-quality public transportation is already available.

Outside those areas, the lower costs and better accessibility that SDVs should bring are expected to drive increased demand for vehicle use. This rebound effect is likely to absorb any road capacity initially freed up by replacing the use of conventional vehicles, potentially exacerbating congestion and emissions. Increased urban sprawl as a result of the large scale use of SDVs is also a concern, due to a potential shift in perception of travel time and use.

The problems associated with nonmonetary use of infrastructure (such as congestion) exist regardless of the presence of SDVs, and there are numerous regulatory options available to handle them. These measures (such as congestion charging) target the price of individual mobility, aiming to internalize the cost of road transportation. In turn, these regulatory options make shared-mobility services more competitive with individual mobility. These measures should be implemented now, to help shift consumer preferences toward shared mobility and improve existing transportation problems.

5.4.5 Regulating SDVs for Urban Shared Mobility

Some key principles must be developed to move forward with data-led regulation of urban shared-mobility services using SDVs. This includes a coordinated, data-led approach for two, related policy areas—SDV-type approval, and mobility service or public-transportation service provision in urban services—and accompanying policy measures. In both the cases, the data-driven processes would solve current disruptive effects of innovative mobility services and provide a sufficiently flexible system to not hinder innovation and systems holding the promise of large-scale improvements in a controlled environment.

Accompanying policy measures might also be necessary to guide technology and business-model implementation toward greater societal good. These measures could include fiscal measures to influence the shift toward more sustainable transportation modes, multi-modal considerations, exercise and health (as many current shuttle systems cover walking distances), land use and parking, pick-up/ drop-off facilities (potentially replacing much of the current parking needs in cities), and traffic management.

In addition, some general principles apply. While future data-led regulatory frameworks should be limited to correcting market failures, they must rely on the most efficient tools, yet also be technology neutral and non-discriminatory. These impacts must be monitored and re-assessed. Regulations must be adaptable, clear, focused, and easy to apply; based on sound economic principles; and inclusive of all involved social groups. Furthermore, division of regulatory responsibility should be adequately divided, particularly when shifting responsibilities from the public to the private sector, as is likely in the context of data-driven processes.

While there are clear benefits of moving toward data-driven governance for urban shared-mobility services, there are many challenges and obstacles to be overcome to enable real-life implementation of such a scheme. Challenges can relate to technical issues, as well as the main involved parties, such as governments, industry, and general society. In the case of government agencies, there could be reluctance to establish new approaches, or a protectionist attitude toward legacy systems. There could also be a lack of data-science competence, concerns over private-sector involvement and responsibilities, or an inability to gain access to necessary data.

In the case of industry, there could be concern over leak of commercial data or issues with competition among actors. There might be concern over leaks of private data for general society, and media framing of large incidents could delay technology uptake. Technical issues to be solved include privacy protection, cyber-security protection, data science developments (particularly blockchain), and road-safety performance of SDVs. Some concrete cases preventing closer cooperation between the public and the private sectors have been investigated as part of the ITF Working Group on Big Data and Open Data. This group is compiling a checklist for setting up data-sharing cooperation between the public and the private sectors. Challenges and obstacles encountered in the workshops and case studies compiled for this group were highly relevant. They included private-sector mistrust in the public sector's capacity to handle data in a sufficiently responsible manner, and other examples requiring careful relationship building. In many cases, these issues prevented cooperation, or limited it to a specific period of time, such as for a research project.

Moving forward with data-led governance of SDVs for urban shared mobility necessitates a coordinated approach for regulating individual vehicles or entire fleets, and the associated transportation service, particularly in terms of road-safety performance and passenger security inside vehicles. Such an approach requires translating current descriptive regulations into quantifiable targets or indicators. In addition, regulator access to data (either mandated or through data-sharing partnerships) could enable a dash-board approach using data analytics, in which instances of crossing pre-defined policy or performance thresholds would be flagged and visualized.

5.5 Conclusion

The challenges and opportunities of establishing data-led governance models for automation and shared mobility have been discussed above. This involved analyzing the related key issues in the transportation and mobility sector, including the sharing economy, disruptive innovation, robot-taxis, current regulatory frameworks, and big data and digitalization.

In summary, the key principles to be followed to successfully move toward the implementation of data-led governance approaches for urban shared-mobility services include a coordinated, data-led approach for the related policy areas of type approval for SDVs, and mobility services. Additional accompanying policy measures include a shift to sustainable modes, multi-modal considerations, exercise and health, land use and parking, pick-up/drop-off facilities, and traffic management. More general points include the fact that regulatory frameworks should:

- be limited to correcting market failures
- rely on efficient tools, but be technology neutral
- be adaptable and monitor impact
- be clear, focused and easy to apply
- be based on sound economic principles
- be inclusive of all social groups
- have an adequate division of regulatory responsibility

The key obstacles and challenges that must be addressed when moving toward implementing data-led governance of urban shared-mobility services can be grouped into those relating to the public sector, the private sector, general society, and technology.

For the public sector, there can be a certain reluctance to establish new approaches, even a somewhat protectionist attitudes toward legacy systems, a lack of data-science competence, concern over private-sector involvement, and an inability to gain access to necessary data sets. From the private-sector perspective, there can be concern over potential leaks of commercial data or competition among actors. From a societal point of view, there can be concern over leaks of private data, or media framing of large incidents. Technical issues must focus on privacy protection, cyber-security protection, data-science developments (particularly the blockchain approach), and road-safety performance.

In conclusion, we are still at an early stage of both the potential mobility revolution (including both SDV and shared mobility discussed here) and a wider move of governments to a digital agenda and economy, leading to many uncertainties going forward. Therefore, it is necessary to study best practices and data from any attempts to implement the kind of data-driven processes that are presented in this chapter.

Moreover, further technology developments, potential societal shifts, and many other factors anticipated through strategic foresight activities must also be taken into account. It is also important to keep in mind that it should not be technology implementation for its own sake. Analyzing the problem must be a starting point, followed by determining if specific technologies, such as urban, automated, shared-mobility services, can be a standalone solution or as part of providing a multi-modal, public-transportation service.

References

Bishop R, Shladover S (2015) EU-US symposium on automated vehicles, white paper I: road transport automation as a public-private enterprise. https://www.ssti.us/wp/wp-content/uploads/2015/10/2015-EU-US-Symposium-White-Paper-I_Public-Private-Enterprise-002.pdf

Botsman R, Rogers R (2011) What's mine is yours: how collaborative consumption is changing the way we live. Collins

Christensen CM (1997) The innovator's dilemma: when new technologies cause great firms to fail. Harvard Business School Press, Cambridge, MA

CPUC (2013) California public utilities commission press release. http://docs.cpuc.ca.gov/PublishedDocs/Published/G000/M077/K132/77132276.PDF

ITF (2015a) CPB report, big data and transport—understanding and assessing options. https://www.itf-oecd.org/sites/default/files/docs/15cpb_bigdata_0.pdf

ITF (2015b) CPB report, urban mobility system upgrade—how shared self-driving cars could change city traffic. https://www.itf-oecd.org/sites/default/files/docs/15cpb_self-drivingcars.pdf

ITF (2016a) CPB report, data-driven transport policy. https://www.itf-oecd.org/sites/default/files/docs/data-driven-transport-policy.pdf

ITF (2016b) CPB report, app-based ride and taxi services: principles for regulation. https://www.itf-oecd.org/sites/default/files/docs/app-ride-taxi-regulation.pdf

ITF (2016c) Case-specific policy analysis report, regulation of for-hire passenger transport: Portugal in international comparison. https://www.itf-oecd.org/sites/default/files/docs/regulation-for-hire-passenger-portugal.pdf

ITF (2017) CPB report, data-led governance of road freight transport—improving compliance. https://www.itf-oecd.org/sites/default/files/docs/data-led-governance-road-freight-transport.pdf

ITF (2018) Roundtable report, cooperative mobility systems and automated driving—summary and conclusions. https://www.itf-oecd.org/sites/default/files/docs/cooperative-mobility-systems-automated-driving-roundtablesummary.pdf

Laney D (2001) 3-D data management: controlling data volume, velocity and variety. META group report. https://blogs.gartner.com/doug-laney/files/2012/01/ad949-3D-Data-Management-Controlling-Data-Volume-Velocity-and-Variety.pdf

McKinsey & Company (2015) Report, urban mobility at a tipping point. https://www.mckinsey.com/business-functions/sustainability-and-resource-productivity/our-insights/urban-mobility-at-a-tipping-point

SAE (2016) Society of automotive engineers: taxonomy and definitions for terms related to driving automation systems for on-road motor vehicles. J3016_201609. https://www.sae.org/standards/content/j3016_201609/

Shaheen S, Cohen A, Zohdy I (2016) Shared mobility: current practices and guiding principles. US Federal Highway Administration, Washington, DC

Shladover S (2017) Connected and automated vehicle systems: introduction and overview. J Intell Transp Syst Technol Plan Oper. https://doi.org/10.1080/15472450.2017.1336053

Sundararajan A (2013) From Zipcar to the sharing economy. Harvard Business Review. https://hbr.org/2013/01/from-zipcar-to-the-sharing-eco

TRB (2017) Transit cooperative research program (TCRP) report 188: shared mobility and the transformation of public transit. http://www.trb.org/Main/Blurbs/174653.aspx

UK Department for Transport (2017) Scoping study into deriving transport benefits from big data and the internet of things in smart cities. Final report. https://assets.publishing.service.gov.uk/government/uploads/system/uploads/attachment_data/file/639027/big-data-transport-scoping-study.pdf

UNECE (1958) Agreement concerning the adoption of harmonized technical United Nations regulations for wheeled vehicles, equipment and parts

UNECE (1998) Global technical regulations for wheeled vehicles, equipment and parts which can be fitted and/or be used on wheeled vehicles

Walker-Smith B (2014) Automated vehicles are probably legal in the United States. 1 Tex A&M Law Rev 411. http://cyberlaw.stanford.edu/publications/automated-vehicles-are-probably-legal-united-states

Tom Voege joined the International Transport Forum (ITF) at the OECD as a policy analyst in 2015. At the ITF, he coordinates work on intelligent transportation systems, including vehicle automation, shared-mobility concepts, and big data in transportation. Prior to joining the ITF, he worked for the United Nations (UN-DESA) as a road and traffic safety expert, following earlier roles in management consulting and academia. He holds an M.Sc. and a Ph.D. in transportation planning and engineering, both from the University of Southampton, UK. He is on the scientific advisory board of the Springer Lecture Notes in mobility and the advisory board of Intelligent Transport magazine. He is also a member of the SAE Committee on Shared and Digital Mobility, and the TRB Committees on Intelligent Transport Systems and Vehicle Highway Automation.

Chapter 6
How Should We Drive Self-driving Vehicles? Anticipation and Collective Imagination in Planning Mobility Futures

Miloš N. Mladenović

> *It is a mistake to confuse progress with a continued movement in one direction.*
>
> Wittgenstein

Abstract Self-driving vehicle (SDV) technology, in its current foundational stage, brings about significant uncertainties, indicating a society-wide disruption. In this context, one cannot disentangle envisioning futures with SDVs from questions of (re)distribution of societal benefits and burdens. Contrastingly, the need for strategies to cope with this disruption has recently been recognized through several planning efforts. Despite their fruitfulness, missing an elaborated understanding of technology as a socio-technical phenomenon remains an underlying challenge. However, reclaiming technological futures as plannable spaces requires understanding that human ends are not well-defined or static, and that technology does not have unstoppable, unquestionable, momentum. At the central point of contention, we argue for replacing the language of unintended consequences with the language of unanticipated consequences. Furthermore, we recognize the threat of anticipation inequality if we solely rely on expert-based practices. To support divergent envisioning efforts, we propose a phase of participatory expansion of technological horizons for (un)desirable futures. To this end, we provide an example from an online discussion platform for reflective engagement about mobility futures with SDVs for a neighborhood in the Helsinki capital region. Finally, we reflect on challenges for inter-organizational learning in coping with contingencies.

Keywords Automated vehicle · Connected vehicle · Technology governance
Transport justice · Transport system planning

M. N. Mladenović (✉)
Spatial Planning and Transportation Engineering, Department
of Built Environment, Aalto University, Espoo, Finland
e-mail: milos.mladenovic@aalto.fi

M. Finger and M. Audouin (eds.), *The Governance of Smart Transportation Systems*,
The Urban Book Series, https://doi.org/10.1007/978-3-319-96526-0_6

6.1 Introduction

The recent convergence of sensing, processing, and communication technology has resulted in a rapid development of self-driving vehicle (SDV) technology (Blyth et al. 2016; Shladover 2017). This development is manifesting itself in a series of pilot tests in urban environments across the world. According to Bloomberg Aspen Initiative on Cities and Autonomous Vehicles, there are 50 cities currently hosting SDV pilot tests or doing so in the near future, while another 23 cities are in the process of long-term surveys of regulatory, planning, and governance issues regarding SDV deployment (Anon., n.d.). However, we must acknowledge that emerging SDV technology remains in the foundational development stage. Although SDV technology is under rapid development, it is still not widely societally accepted, and there are multiple competing premises behind relevant issues (e.g., traffic safety, climate change mitigation), as well as visions of futures. Consequently, SDV development has not yet reached a major milestone in its technological trajectory in which SDVs are embedded into the everyday public sphere, even at one urban location.

Understanding that SDV technology is in a foundational stage is just one essential premise required for a deeper understanding of technology as a socio-technical phenomenon and for its governance. One starting premise is that governance of smart mobility technology must relate to the notion of ensuring and enhancing public value (Docherty et al. 2017; Pangbourne et al. 2018). Moreover, another premise is that transportation planning and policy-making face deep uncertainty, to which they must respond with an envisioning approach that confronts biases and uses flexible methods (Lyons and Davidson 2016). However, what we aim to show in this chapter is that SDV governance must expand its conceptualization of technology as a socio-technical phenomenon. With a pragmatic aim to inform the development of strategies and instruments to cope with societal disruption in relation to SDV technology, we intend to deepen the understanding of technology as an instrument of governance in itself. Throughout the chapter, we use the term planning (rather than the term governance) as a broader term encompassing decision-making processes about desirable, possible futures (Connell 2009). In order to develop strategies and instruments capable of imagining opportunities and undesirable consequences, we must recognize that there is previous societal experience with managing and planning emerging technologies, so we do not have to invent the wheel again. With this in mind, we will build the argument by extensively leaning on key conceptualizations of technology existing in theories of science and technology studies (STSs), as well as in the philosophy of technology. Empirical material used in this chapter will only serve the purpose of exemplifying and adding depth to the existing, experiential dimensions of inquiry about SDVs as an emerging technology.

Highlighting the need for institutional learning in planning emerging technologies, we should understand that these technologies are often perceived as an opportunity for reframing policy-making toward more active engagement in shaping open-ended technological systems (Rayner 2004; JafariNaimi 2018). Conceptualizing governance as social learning is particularly pertinent to emerging technologies. In the case of SDVs, we already see implications for data sharing between companies to enable that learning process (Stilgoe 2017). In a similar manner, we can build upon the knowledge from theories of anticipatory and reflexive governance (Voss et al. 2006; Quay 2010; Guston 2014), as well as responsible innovation (Owen et al. 2012; Stilgoe et al. 2013). These existing planning frameworks can contribute to the underlying need to understand systems with which technologies may coevolve, as well as explicate reasons for how and why innovation happens.

At this point, it is important to recognize that our argument is localized in the Nordic context, focused on Finland, which is currently undergoing rapid urbanization, bringing a range of opportunities and challenges. We recognize that recommendations laid out here are taking into account some distinct features of Finnish society, such as strong know-how in technology development, a knowledge-intensive economy, a good innovation environment, and high technological knowledge across generations. Moreover, Finnish society has a value system emphasizing citizens' rights, intellectual autonomy, equality, self-expression, creativity, cooperation, respect for nature, social cohesion and harmony, reduced societal hierarchy, a preference for trust-based social norms, and transparent policy-making processes (Haldor et al. 2002). By focusing on urban environments, this discussion excludes reflections on particular aspects of the transportation sector, which are long-distance travel and freight transportation. Furthermore, we recognize that there are previously identified path dependencies in Finnish transportation innovation policy (Upham et al. 2013), which are outside the scope for this chapter.

The following section will elaborate on a wide range of uncertainty factors related to SDV technology deployment, pointing toward a society-wide disruption and an inevitability of the underlying ethical challenge. The third section of this chapter will reflect on the ongoing technology development and planning efforts, explicating some of the underlying assumptions. In order to deepen the conceptualization of SDV technology, the fourth section will elaborate on the meaning of technology and the need to shift discussion from envisioning uncertain effects to unanticipated ones. Having the elaborated understanding of technology as a phenomenon in mind, the fifth section will provide an example of an online discussion forum established in Finland as an example of engaged collective deliberation about technological futures. The sixth section will reflect on the implications from this deliberation exercise. The seventh section will reflect on further challenges for inter-organizational learning in technological planning, while the chapter itself ends with a brief conclusion section.

6.2 Disruption from a Wide Range of Interdependent Contingencies

As mentioned in the introduction, SDV technology is currently in its foundational stage, emerging in several versions, including passenger vehicles, minibus shuttles, buses, and freight trucks. All of these types of vehicles have a range of different self-driving features. Ultimately, the expected SDV function is replacement of parts or the whole driving task (Banks et al. 2014). These current multiple versions of technology are inevitably related to multiple possible futures. In addition, a range of interdependencies between societal aspects shaped by SDV technology inevitably raise questions about a wide range of uncertain consequences. In particular, fundamental challenges appear when trying to imagine SDVs as technology embedded in the urban built environment. One fundamental challenge is a difference between the dynamics of technological and built-environment development, with the latter often having a relatively slower development cycle to the former. Therefore, consequences of SDVs embedded in built environments are much harder to estimate in comparison to less pervasive, less mundane, technologies.

In response to the general sense of uncertainty, research efforts inform us that we can expect a range of potential changes, primarily related to transportation systems (Lutin et al. 2013; Thomopoulos and Givoni 2015; Milakis et al. 2017a, b). A rich Web of interdependent factors and uncertainties usually starts with effects on whole built environments, as well as other technologies and services (Schiller 2016; Zakharenko 2016). For example, either by relying upon previous research or through structured reflection, SDV technology might pertain to changes in link and network capacity; overall road network layout; street surface and cross-sectional design; integration with other transport modes, especially in relation to station or stop design; parking planning; urban freight planning; traveler information technology, and general data management architecture; vehicle-charging infrastructure; algorithms for demand-responsiveness; area and time limitations; or user interfaces for mobility services (Fagnant and Kockelman 2015; Talebpour and Mahmassani 2016; Chen et al. 2016). With these example in mind, SDV technology is often interpreted as promising several potential benefits, including reduced road accidents; reduced GHG emissions; energy savings; congestion avoidance; parking savings; new on-demand mobility services and service integration; mobility for elderly, children, and the disabled; in-vehicle productivity; logistics productivity; and overall export potential.

In addition to potential changes in the built environment, technologies and services, SDV technology also pertains to changes in organizations and policies (Gettman et al. 2017; Vellinga 2017; Schellekens 2015). First, SDV technology might relate to changes in required knowledge and planning practices, policy coordination practices, decision processes about infrastructure investments, models for distributing operating costs and ownership, and mobility marketing and education practices. Such changes are relevant for municipalities, planning organizations, road-maintenance authorities, and engineering consulting companies. Second, by being related to changes in mobility patterns and pricing schemes, organizations such as public transportation agencies, parking agencies, taxi companies, transportation network companies (TNCs), freight companies, retailers, air and rail operators, auto repair companies, hotels and rest areas, real-estate companies, and telecommunication companies will also be affected. Moreover, SDVs might relate to changes in taxation and insurance regulation, as well as data management and ownership regulation. This could make them pertinent for organizational disruption regarding insurance companies, law firms, enforcement agencies, emergency healthcare providers, and energy providers. Many other contingencies can be associated, with respect to work regulation, public services funding schemes, or job loss and shift.

Finally, and perhaps most importantly, SDV technology has the potential for changing a range of peoples' everyday activities (König and Neumayr 2017; Harper et al. 2016; Auld et al. 2017). For example, SDV technology might pertain to changes in use of non-driving time, acceptable walking distances, year-round patterns of mode choice and travel distances, travel reliability and speed expectations, willingness to pay for services. These potential changes are even more pertinent, taking into account development in mobility-on-demand services (Jokinen et al. 2017; Pangbourne et al. 2018). Further reflection might lead to a conclusion that SDV technology has potential for truly society-wide effects, since the disruption also relates to societal values and norms. Pushing the thought experiment even further, further changes can be imagined in a range of societal norms, such as trust and perception of safety and security in connection to interaction between inside and outside of the vehicle. Moreover, changes might pertain to a shift in peoples' perceptions of responsibilities for critical events and accidents; changes in the concepts of least-advantaged and dignity, such as in the case of dedicated SDV use by the elderly or children; and changes in the societal value of a physically active lifestyle.

We will not provide an extensive list of affected factors and their often-unknown interdependencies. Nonetheless, the above examples of significant uncertainties should be sufficient to point toward potential society-wide disruption that is certainly beyond the narrow definition of transportation as one, clearly outlined, societal sector. Metaphorically speaking, just as the personal computer did not only replace the typewriter, but overhauled our entire society, a similar scale of societal change is expected from SDV technology. In addition to recognizing the extent of numerous factors that might be affected, we must acknowledge that mobility, as a centerpiece of everyday life, has ethical dimensions. In fact, we must recognize that technology not only enhances mobility, but also shapes it (Bergmann and Sager 2008). One seminal STS argument is that technology frames rules as much as laws do, with power to order society by enabling and constraining basic human rights and responsibilities (Winner 1977). Similarly, we must remember that technologies do not solely bring about societal benefits, but also create problems of their own, often by locking us into undesirable futures (Latour and Venn 2002). Consequently, we cannot disentangle questions of socio-technical transition from (re)distribution of benefits and burdens across present and future societal groups, as well as the threat of reproducing existing forms of inequality (Blyth et al. 2016; Mladenovic et al. 2017). With this ethical dimension in mind, planning strategies and instruments must cope with the fact that we need reflection on societally (un)desirable futures, not solely technically possible ones.

6.3 Emerging Planning Efforts and a Threat of a Faustian Bargain

As our focus is on providing lessons for planning mobility futures, we must now acknowledge a range of actants in the current technology development processes. These development processes still rely on division of labor, in which companies are mainly in charge of technical development, while the public sector mainly gets involved in ex post assessment, regulation, and implementation. In this context of asymmetry between development and implementation, technology development processes are often object-centered and overwhelmingly focused on time-to-market performance measures, so do not take into account the complete implementation context of the artifact (Stilgoe 2017). Furthermore, the resulting concentration of decision-making power in a limited number of organizations (re)produces a power imbalance between those who develop and implement new technology and those whose lives will be reshaped.

When we turn toward public institutions in relation to the technology development processes, we must recognize that emerging technologies, such as SDVs, typically face the challenge of institutional void (Hajer 2003) and organized irresponsibility (Beck 1992). Basically, this means that none of the current institutions has a full understanding or control of undesirable consequences associated with SDVs. In turn, the resulting distributed responsibility for transition management and technological development limits individual and institutional accountability. Contrastingly, the need for strategies and instruments to cope with disruption in response to contingencies has recently been recognized through several national and city planning efforts (Fagnant and Kockelman 2015; Guerra 2016; Williams et al. 2017; Anon. 2016; Walker and Marchau 2017; Zmud et al. 2017). Although focused primarily on the transportation sector, these efforts have aimed to provide recommendations for planning organizations beyond the limitations of conventional infrastructure programming practices anchored in forecasting-based travel behavior modeling using aggregate trip-based models, and project appraisal, using such tools as cost-benefit, environmental impact, or even multi-criteria analysis. In particular, very recent, limited efforts have utilized agent-based, system dynamics and accessibility modeling for stimulating reflections about desirable futures with SDVs (Fagnant and Kockelman 2014; Nieuwenhuijsen et al. 2018; Meyer et al. 2017). Moreover, state-of-the-art planning methods used for reflection about futures with SDVs have centered on a vision-based approach, building upon insights from future studies by operationalizing such foresight methods as scenario planning and backcasting (Fraedrich et al. 2015; Milakis et al. 2017a, b).

These efforts must also be contrasted against a recent trend in surveys of public acceptance of SDV technology (Becker and Axhausen 2017). We could argue that methodological limitations of surveys require constraining the number of technological options. The public is only asked if these particular versions of SDV technology are acceptable. These surveys do provide a range of findings and points for further reflection. However, we must be aware of the potential threats of discounting undesirable consequences and neglecting technological dynamics. On the one hand, these threats might stem from a lack of reflexivity about contingencies themselves (Latour and Venn 2002). On the other hand, keeping in mind previously mentioned methods and their use in actual planning processes, we might still lack the capacity to imagine breaking existing technological path dependencies (Mladenovic et al. 2016), as well as avoiding potentially irreversible technological lock-ins. Simply put, current discourse misses an opportunity for more divergent reflection on all possible, desirable technological options. For example, even simple reasoning about a vast diversity of current vehicle types would imply the possibility and desire for a large number of different SDV versions. In practice, such a variety could be developed by accounting for particular geographical, infrastructural, institutional, and certainly societal contexts. Here, an old dilemma emerges. On the one side, we are unable to estimate the changes from emerging technology until the technology is fully formed and embedded in society. On the other hand, changing a technological development trajectory is very difficult once the technology is fully formed (Collingridge 1980). In this situation, there is a threat of a Faustian bargain that aims for societal benefits but only under conditions of unequal bargaining power and irreducible uncertainty.

6.4 The Meaning of Technology and a Shift from Uncertain to Unanticipated Consequences

These previous planning-related efforts provide a range of recommendations and valuable developments, identifying a spectrum of societal and organizational factors for which transition strategies must account. Despite these fruitful efforts, at least one underlying challenge remains—lack of an elaborated understanding of technology as a phenomenon. In particular, it is important to recognize that technology is never static, but always changes over time (Jasanoff 2016). On the other hand, as humans are profoundly technological beings, changes in technology also imply changes in defining what it means to be human, as the fuzzy boundary between the two is constantly reshaped (Pacey 1999; Jasanoff 2016). Simply put, by asking what a technology is supposed "to do", such as in the case of SDVs often argued "to remove parts or the whole driving task from human", we cannot avoid to simultaneously ask what will "remain" for the human "to do" at the same time.

Moreover, removing parts of the driving task not only leaves the remainder of tasks inside the SDV for the human, but also generates new tasks in response to the emergence of SDVs in our built environments and institutions. With SDV technology in mind, this reshaping has the potential to affect the very meaning of our everyday lives. By avoiding bringing forth the dynamics and reshaping of human meaning in reflecting about technological development, there is a simultaneous danger of not accounting for the full extent of ethical implications. Furthermore, as already identified in the realm of territorial governance (Metzger et al. 2015), there is a threat that some fundamental issues will be depoliticized or even rendered invisible in technology development processes, especially in the foundational stage.

At this foundational technological development stage, it is important to recognize two essential premises for developing planning frameworks. First, if we are to reclaim technological futures as plannable space, we must tackle the challenge of perceiving human ends as well-defined and static, to which we only need to provide technological means (Jasanoff 2003, 2016). Making this assumption about the nature of human beings carries the risk of disregarding evolving relationships between us and our technology, as well as often irrevocable changes, not solely to the built environment around us, but also to our fundamental values and norms. Second, a challenge of technological determinism in the foundational stage of emerging technology might lead us to conclude that technology has unstoppable momentum, reshaping society to fit to its demands (Jasanoff 2016). Simply put, SDVs, just as climate change, are inevitable. This assumption is often underlined with the fact that technological pathways are less linear and more difficult to predict, due to their increasing complexity. As a result of this rapidly increasing nonlinearity, our understanding of technological development trajectories usually happens in hindsight. However, for the more difficult exercise of foresight or envisioning, we must raise the question of organizations' or even society's imagination capacity.

Starting from understanding the span of uncertainties, we have raised the question of desirable and undesirable consequences. However, we must dedicate special effort to avoid falling into the trap of reflecting about unintended consequences. The very language of unintended consequences implies that it is not possible to think about the full range of undesirable consequences. By removing the assumption of possibly thinking about undesired consequences, technology creators give away their power to question if alternatives to inbuilt societal structures are at all imaginable. As an example, let's briefly reflect about a major initiator of the modern automobility regime, Henry Ford (Urry 2004). In hindsight, one could wonder how Ford could foresee drastic consequences for the climate and society that his creation of the passenger car artifact would bring about after many decades. After all, Ford intended to achieve a set of desirable consequences by unlocking immense possibilities for individuals of all social classes. Perhaps we can even give Ford the benefit of the doubt that he could not have foreseen all the undesired consequences of the automobility regime that are currently devastating our societies across the globe. Nonetheless, we must recognize the challenge that we often reflect on desired outcomes in the future, while discounting harms as more speculative

(Jasanoff 2016). Moreover, we must recognize that, as far as SDVs are concerned, we are no longer in the early twentieth century. We now have a greater amount of societal knowledge when we talk about emerging and disruptive technologies. On the one side, the least we can do is learn from the development trajectory of passenger automobiles. On the other hand, we can reflect upon the development trajectory of personal computers as an equally disruptive technology present in our everyday lives. Therefore, if we move away from the language of unintended consequences to embrace that we are solely talking about unanticipated consequences, we can build upon the societal knowledge to empower our collective imagination.

6.5 Exercise in Collective Imagination of Mobility Futures

By recognizing the need to shift planning toward examining unanticipated consequences, we must recognize the threat of anticipation inequality if we only rely on expert-based practices. To support planning and policy-making in dealing with technological disruption, we must enable a phase of participatory expansion of the technological horizons of desirable futures (Blyth et al. 2016). In addition, reflections on socio-technical transition require a discussion about (re)distribution of benefits and burdens across societal groups, significant uncertainties, and threats of reproducing forms of inequality. Following an open, responsible, innovation approach and a strong culture of citizen participation in Finland, we organized an online forum for reflective engagement. The goal of this online forum was to enable discussion about mobility futures for Otaniemi, a neighborhood in the Helsinki capital region, where the main campus of Aalto University is situated. In line with thinking about the SDV artifact and its context, a description of the vision was posted in the introduction to the discussion forum. The vision was called self-driving urban area (SUA) and included:

- Shared self-driving electric shuttles, with fixed and on-demand routes;
- Proximity to high-capacity transportation nodes;
- Integrated land use, and transportation and energy infrastructure planning;
- Emphasis on street design for walking and biking;
- Restricting car access by managing parking;
- Public transportation and pricing policy;

The participants were encouraged to share their comments about the vision for Otaniemi, including SDV shuttles and a comprehensive built-environment plan, as well as to further engage with each other's comments. A pilot trial of a self-driving shuttle was simultaneously organized in the Otaniemi area. The shuttle was fully electric, with dimensions of 3.9/2.0/2.8 m (length/width/height), with a passenger capacity of 12 (six seated, six standing). For the discussion platform itself, the research team used an existing online community, Hubchat, which was easily accessible using a social network or email sign-in. The discussion community was available through a browser and app-based interface, as in Fig. 6.1.

More than 70 members signed up for the community for three months in 2016, resulting in dozens of discussion posts. Members' comments discussed scope and desirability in relation to SDV technology and SUA concept. The platform attracted both male and female participants, including Finnish and other nationalities. The community did not have moderators or any participation-activation techniques, but still succeeded in maintaining a solid level of activity. The community featured participants with multiple posts and comments, and followers with less activity were mainly active by asking questions. However, there were also users who mainly observed and did not get directly involved in discussions. Open-ended discussions enabled a wide range of topics to appear, enabling a snowball effect, in which discussions continued from previous points. Consequently, users provided a lengthy narrative that was rich enough to gain insight into their perspectives and relationships with the technology. Moreover, users commented on each other's statements, effectively building a space for honest and messy debate. Following are some excerpts of the written narratives that aimed to uncover additional dimensions for discussion or provide evidence for an STS knowledge base used in formulating the argumentation of this chapter. Two following comments depict contradicting impressions of the atmosphere inside the self-driving shuttle piloted in Otaniemi, raising questions of safety and privacy.

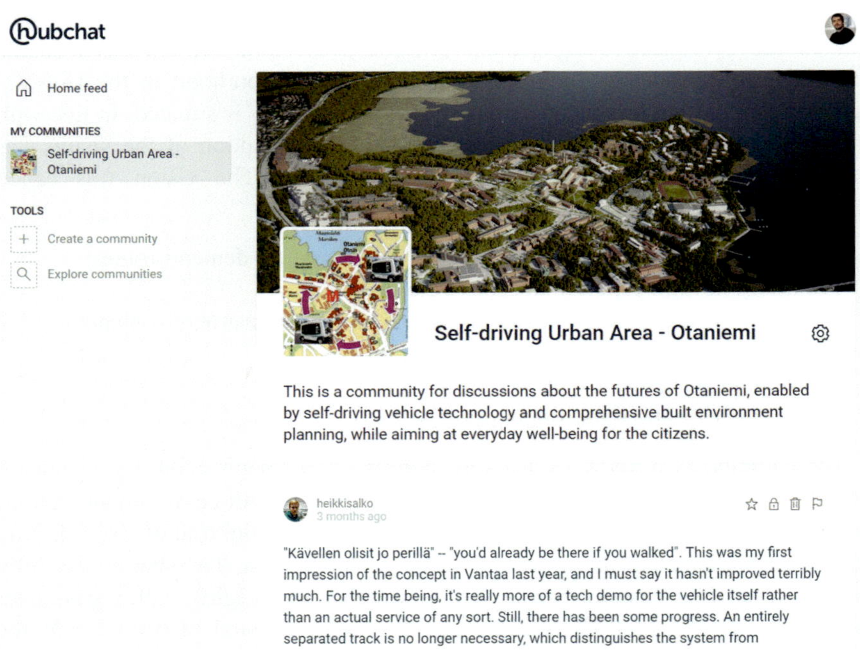

Fig. 6.1 Snapshot from self-driving urban area Otaniemi discussion forum

> I was quite excited to try out the self-driving shuttle bus in Otaniemi last Friday. Even though I believe that the self-driving cars are or at least will be safer than people on the road, I was a bit unsure about the safety.
>
> The design of the self-driving shuttles reminds me at a cabin of a cable railway and sitting inside feels very different compared to a standard bus. In my eyes the largest difference was the atmosphere, which was more private and gave me rather a feeling of holiday than of everyday life.

Several discussions revolved around the question of shuttle speed, which was limited to 11 km/h for the pilot, even if the vehicle could go up to 40 km/h. The factor of speed was interpreted in relation to other aspects, such as travel time and safety while traveling in Otaniemi.

> With the current speed, I think the self-driving bus could only be a competitive mode on a cold winter day when everybody avoids walking and only if you are not in a hurry.
>
> Coming back to this now, yes, I would say speed is important. The only advantage of slow speeds that I can see is the ability to potentially share paths with pedestrians with relative safety.

One example of an extensive debate with conflicting opinions was around the notion if the self-driving shuttle competed with, or complemented walking and cycling. Participants imagined a range of circumstances, and used cases, for a self-driving shuttle in Otaniemi, recognizing that there might also be particular groups of citizens that could benefit from SDVs. In sum, these and similar discussions inevitably led participants to raise questions about citizens' rights and equality in relation to mobility.

> In my opinion, they [SDVs] could support the walking mode, allowing pedestrians, e.g. coming from shop-ping carrying grocery bags, to take a shuttle bus to ride over a short distance before continuing walking to their final destination.
>
> Cycling as a transport mode in Otaniemi area is superior compared to any other mode. The viability and convenience of the SUA shuttle is greatly hindered because of this.
>
> Cycling is indeed far more superior in Otaniemi and will be so for a long time. But cycling isn't an option for example for people with special needs, visitors and others not owning a bike or people that have traveled to Otaniemi by other means of transport and need to travel within the Otaniemi area. The SUA shuttle can serve these groups of people better than cycling.
>
> Now, there are reasons I might still opt for a shuttle. In pouring rain it would be preferable to travel in a covered pod and arrive relatively dry. But this requires that I don't have to stand waiting in said rain for very long, so short wait times and/or shelters are necessary. I also might be carrying something big and heavy. Then it needs to be possible to ensure that my cargo fits inside the vehicle with the other passengers. I could also be injured or old and have difficulties walking.

6.6 Implications from Collective Imagination Exercise

As shown above, engaging citizens to co-creatively imagine how SDVs could be integrated with built-environment planning is an essential aspect of the proposed envisioning approach, as processes for coping with disruption should account early on for citizens' perspectives. Although the online discussion platform was a successful instrument for engaging citizens in the debate, there are a couple of limitations that must be addressed. First, the platform mainly attracted students, particularly those studying engineering and technology. This was expected, considering that these students are usually interested in emerging technologies and have shown great interest in the SDV shuttle pilot program. Future deployments of these platforms should use mechanisms to expand the user base. Second, since English was used as the platform language, it might have been a limiting factor for wider participation. Future deployments should also take language into account, especially when considering older users. Third, the fact that SDV shuttle used in the pilot program was shared, it framed the discussion toward aspects of shared vehicles. For example, several participants raised questions of privacy and security, which might not be brought up for privately owned SDVs.

Snapshots of issues raised in the community point toward the need to recognize the difficult discipline of deliberation in equalizing opportunities for anticipation beyond expert-based mechanisms. Such open-ended discussion allowed for explicating a unique assembly of Finnish social and cultural norms and values. Explicating these values and norms in the foundational stage of technological development can hinder or help comprehensive technological utilization. Returning to the relationship between possible—desirable—anticipated futures, we must recognize the need for spaces for contestation about the desirable and anticipated. The open-ended discussion also enabled the emergence of disagreements and points of conflicting opinion. Such conflicting visions of the future may be inevitable, as different individuals might have different rankings of importance for socio-technical factors. Nonetheless, such controversies might be useful as an informal way of assessing technology, enabling social learning by not immediately seeking consensus (Rip 1986). Thus, our capacity for collective imagination should not be hindered with a premature focus on consensus-seeking. On the contrary, democratization of technological development processes has an enormous capacity for avoiding displacement of the political debate about the inevitable ethical underpinnings of the disruptive SDV technology. Consequently, mechanisms for acknowledging the right of conflicting visions to emerge, and be deliberatively resolved, must be found. Envisioning elaborate scenarios of lives in the future must also recognize that thinking about undesirable urban futures can help in reflection about unanticipated consequences. Conclusively, one potential planning approach is to start from identifying undesirable visions of the future and building consensus about desirable visions through that.

As the process of planning technological transition is a future-oriented activity, it must always balance between a deterministic approach (extrapolation), in which certain existing trends continue as they are, and an envisioning approach that creates new visions of technology. Here, we must highlight that forum participants also acknowledged that, despite being primarily a mobility technology, SDVs might significantly affect individuals' household, work, leisure, and other daily activities, in addition to commuting patterns. Therefore, dealing with technological disruption will require elaborating visions of daily activities in the future, not solely visions of cities or regions. Moreover, these visions must account for various groups, as forum participants discussed. For example, it will be important to focus on activities of the elderly and disabled by evaluating consequences to improving their functional and working ability, children's independent mobility in urban environments, and questions such as increased workplace flexibility location and emergence of new forms of employment. Only after elaborating the visions of everyday life, can the focus shift to thinking about systemic, emergent effects at the city scale.

Considering the discussion forum presented above, there are considerations for further development that could be identified for different phases of the online community's lifetime. For example, moderators could follow the discussions in real time to identify if there is a potential for interference, as well as subtopics that may require a separate subpage within the same discussion community. The separation into subtopics could be enhanced with placeholder discussions, as well as by allowing the community to vote on the rules for managing content. Future deployments could also consider location or event-based activation through questionnaires or pop-up messages, especially for smart-phone applications. The application would follow the principles of experience sampling and could be well-combined with future SDV technology pilot programs. For example, certain features could be enabled or disabled if the user is located within a certain range of the pilot test area. Moreover, different activation techniques could be used in combination with techniques for identifying leaders, followers, and observers.

6.7 Further Challenges for Inter-organizational Learning in Planning (for) SDVs

Building upon our previous arguments, this section provides some reflections about challenges ahead in developing technology-planning frameworks, while recognizing that our legal and political sophistication must evolve along with our technologies. Therefore, reflections are presented under the assumption of long-term organizational learning practices and knowledge management, keeping in mind an outline of preferred planning processes. First, considering the previous points, one must think about both the desirable features and the technically feasible features of SDV technology. This continues into focusing on both functional and technical requirements, which consequently frame international standardization of technological components. Some overlap of functional and technical requirements is to be expected across different cities and regions and has the potential to guide international standardization efforts. However, some divergence of function and, consequently, technical requirements should be expected and accepted as something to be tackled at the national or international level. Thinking about SDV-based technological transition at the city and regional levels is important from at least two standpoints. One is that there are significant enough differences between urban regions across the world for them not to adopt just one version of SDV technology. A simple example would be differences in climate conditions. The second is the question of scale and technological development processes, since establishing the requirements at the national level might be too general for actively guiding the technical development processes. At the same time, establishing them at the level of neighborhood might be too specific for enabling larger standardization efforts.

In recognizing that our current, mainstream planning methods are not well-equipped to cope with the level of societal disruption and uncertainty that stems from SDV technology, we should expect challenges in developing appropriate planning methods. In particular, in addition to advancing the agility of planning processes, decision-support processes, tools, and performance measures should also be elaborated. Here, we should consider utilizing qualitative methods combined with existing, advanced, quantitative methods in transportation planning and policy-making, such as agent-based and system-dynamics modeling. However, such advances in decision-support methods have the opportunity to be organized into a participatory, planning support system. Such a system can be a flexible combination of digital and non-digital tools, which can utilize collaborative arrangements and visualization capabilities in the form of a decision theater setting. Nonetheless, even when developing some of these methods, we must recognize the fact that some consequences cannot be assessed in a quantitative manner during the foundational development stage.

As one additional aspect, we should reflect about challenges for policy coordination, keeping in mind the current institutional constellation. Considering the complexity of SDV technology and its consequences at the societal level, leaving the technological transition to global market forces will not only result in Finland underutilizing this technology, but potentially have very detrimental consequences on the whole society. Moreover, even a glimpse at factors and stakeholders relevant for this technological transition informs us that SDVs are not solely a question for the transportation sector. Therefore, the public sector must simultaneously focus on several aspects for developing optimal transition plans. For example, coherence must be introduced into the policy portfolio, as opposed to the current practice of silo-based policies and local activities. Moreover, there is a need for continuing developing integrated planning approaches for land use, housing, transportation, service structures, and operating conditions for business (MALPE). In addition to vertical, horizontal, and timing coordination of policies, the public sector must consider a range of instruments, from taxation and infrastructure investments, to regulation and incentives for flexible work practices. In order to enable these changes in public sector institutions and operational methods, the Finnish public sector must further understand the mechanisms for effective policy learning and development of strategic agility, as well as for improving conditions for institutional decision-making transparency for successfully anticipating and managing contingencies.

In addition to selecting actions, we must recognize that framing and shifting legal and financial incentives that are established for R&D and customizing technological development will also imply statements about the transition timing. As time is an important question in any technological transition, and even more so with this level of potential societal disruption, we should avoid falling into the trap of overemphasizing that SDV technology should be deployed as soon as possible. Strategizing requires that we liberate ourselves from the often-used assumption in foresight methods that roadmaps are built with a certain end date in mind at the start of the planning process. Instead, the transition end date should be the outcome of the divergent and agonistic envisioning process, taking into account both the desired end state and the means for achieving it. To this end, it is important to remember the strategic importance of the relevant decisions over the long term. Therefore, the question at hand is not solely just the duration of the transition itself, but also the duration of its effects. Reflecting about the passenger car in the twentieth century reminds us that some decisions or actions that made sense in the short term, and could be delayed for a few years, might not have such negative or difficult-to-modify effects for decades to come.

Considering the scale of the challenge at hand, an inevitable set of questions rises in relation to organizational networks and associated mechanisms for genuine accountability and transparency in technological development processes. In the case of Finland, SDV technology directly relates to the country's priorities in developing intelligent systems and services in urban environments, intelligent energy systems, the Internet of things, as well as social media and mobile Internet technology. Therefore, the question remains: Can Finland create competence-based growth, using its strong know-how in sensing technology, mobile devices and applications, accompanied by excellent telecommunication infrastructure? Moreover, can Finland build upon its value-added and knowledge-intensive economy while exploiting its reputation for high-quality products, a strong innovation environment, and the high-tech knowledge level of younger generations? Here, the public sector must engage with exceptional leaders and pioneers, with cross-generational thinking to skillfully connect and mobilize complex networks of actors across system levels and organizational boundaries in establishing potential public–private (people) partnerships. In addition to the responsibility for the public sector, there is a need for changes in business model innovations and roles of organizational strategy. There is a need to leverage SDV technology by developing supportive innovation networks and a national-level value-creation ecosystem of companies, especially by highlighting the need for engaging different user, producer, and platform actors to address the true needs for mobility solutions. Ultimately, experimentation with new institutional forms through collaborative organizational networks may open up the constellation of current roles and responsibilities for change agents, and perhaps nudge governance to step outside of artificial silos and focus on everyday life, with a shift toward a phenomena-based approach.

6.8 Conclusion

Starting from the notion that SDV technology is currently in its foundational stage, this chapter has aimed for a deeper conceptualization of technology as a socio-technical phenomenon, in order to reflect on the instruments for achieving public value while successfully coping with deep uncertainty. The reflection has been provided in the context of a Nordic democracy, assuming planning and technological development as a social learning process. In this context, we tried to highlight that we cannot disentangle envisioning futures with SDVs from questions of (re)distribution of societal benefits and burdens. Nonetheless, we recognize that there are several recent planning efforts that started with the need to cope with disruption. Despite their foresight and systems approach, an elaborated understanding of technology as a socio-technical phenomenon remains a missing component, manifesting as a lack of critical reflection on all possible and desirable technological futures.

Contrary to previous efforts, we argue that reclaiming technological futures as plannable space requires understanding that human ends are not well-defined and static, to which we just need to provide technological means. Moreover, we argue that technological determinism might lead us to conclude that technology is unstoppably reshaping society to fit its demands. At the central point of contention, we highlighted the need to replace the language of unintended consequences with the language of unanticipated consequences. In relation to this need, we aimed to recognize the threat of inequality of anticipation if we solely rely on expert-based practices. Moving away from only emphasizing expert knowledge, we propose a phase of participatory expansion of technological horizons for (un)desirable futures that can support agonistic envisioning efforts. Consequently, we could avoid the pitfall of rushing headfirst toward the one new artifact by enabling our collective capacity for imagination and emancipation. To this end, we organized an online discussion platform for reflectively engaging about mobility futures with SDVs in a neighborhood in the Helsinki Capital Region, Finland. Discussions during three months in the fall of 2016 provided a range of lessons on the demanding discipline of deliberation about contested futures. Finally, a set of reflections on challenges for inter-organizational learning included such questions as standardization practices, policy coordination and timing, and experimentation with organizational constellations.

At the core of our argument is the need for a conceptual shift in viewing SDV technology. Such a shift must recognize a wide range of factors and their interdependencies, due to the scale of potential disruption stemming from SDV technology. In general terms, these factors span from infrastructure and technology, services, organizations and policies, public behavior and attitudes, and societal norms and values. Therefore, dealing with technological transition centered on some version of SDVs cannot remain techno-centric, as we have already learned about the passenger car transition in the twentieth century. For example, dealing with technological disruption will certainly require accounting for path dependencies in the existing infrastructural systems that can prevent or enable SDV technology diffusion. However, the question of technological transition also relates to such questions as citizens' rights, equality, respect for nature, social cohesion, and privacy. Since the SDV technology is still in its foundational stage, now is the critical stage to account for discussing essential societal values. Neglecting to take into account this side of the coin, which involves a unique combination of social and cultural norms and values in each city or region, means neglecting an important aspect for hindering or helping comprehensive technological utilization.

Finally, we would like to highlight that societies articulate their hopes, dreams, and desires through technology, while also making material instruments for accomplishing them. SDV technology may be ultimately framed as a lens for rethinking conventional approaches for addressing contemporary social, environmental, and economic challenges in urban environments. One point to help reflect about the future of SDV technology is to not focus on the technology itself, but to use it as a sort of a prism for addressing grand societal challenges. As highlighted in our examples, this can be done by offering fresh ideas for the future of the economic and working life, enabling regeneration of the public sector, or improving well-being across societies. In doing so, technological transition will not solely address existing problems, but also avoid creating new problems. Although this effort is directed toward all planning organizations, a particular point is in order for transportation and mobility researchers. Transportation and mobility, as a broad research field, has been multidisciplinary since its origins. Perhaps it is time for a transdisciplinary focus on elaborating the meaning of transportation technologies. For this effort, we may need greater engagement with technology studies scholars in the wider fields of future studies, design, sociology and philosophy, as well as building upon the lessons of such theories as actor-network and activity theory. Furthermore, these efforts might contribute to further conceptual refinement of the meaning of everyday technology. Ultimately, transportation and mobility researchers must face a challenging question: Will our community recognize this historical opportunity to redefine the meaning of technological progress and the consequent role that collective imagination of (un)anticipated futures has in such a reflection?

Acknowledgements This research has been supported with funding from the Academy of Finland, through Strategic Research Council project BEMINE (No. 303538), and by the Climate-KIC, through SELF-DRIVEN Areas project.

References

Anon. (n.d.) Bloomberg aspen initiative on cities and autonomous vehicles. Available at: https://avsincities.bloomberg.org. Accessed 1 Dec 2017

Anon. (2016) Federal automated vehicles policy. Available at: https://www.transportation.gov/sites/dot.gov/files/docs/AVpolicyguidancePDF.pdf. Accessed 26 Nov 2017

Auld J, Sokolov V, Stephens TS (2017) Analysis of the effects of connected-automated vehicle technologies on travel demand. Transp Res Rec J Transp Res Board 2625:1–8
Banks VA, Stanton NA, Harvey C (2014) Sub-systems on the road to vehicle automation: hands and feet free but not mind free driving. Saf Sci 62:505–514
Beck U (1992) Risk society: towards a new modernity. Sage, London
Becker F, Axhausen KW (2017) Literature review on surveys investigating the acceptance of automated vehicles. Transportation 44(6):1293–1306
Bergmann S, Sager T (2008) The ethics of mobilities: rethinking place, exclusion, freedom and environment. Ashgate, Farnham
Blyth PL, Mladenovic MN, Nardi BA et al (2016) Expanding the design horizon for self-driving vehicles: distributing benefits and burdens. IEEE Technol Soc Mag 35(3):44–49
Haldor B, Myklebust S, Myrvang C, Sejersted F (2002) The democratic challenge to capitalism: management and democracy in the Nordic Countries, 1st ed. Copenhagen Business School Press
Chen TD, Kockelman KM, Hanna JP (2016) Operations of a shared, autonomous, electric vehicle fleet: implications of vehicle & charging infrastructure decisions. Transp Res Part A Policy Pract 94:243–254
Collingridge D (1980) The social control of technology. Open University Press, London
Connell DJ (2009) Planning and its orientation to the future. Int Plan Stud 14(1):85–98
Docherty I, Marsden G, Anable J (2017) The governance of smart mobility. Transp Res Part A Policy Pract. http://dx.doi.org/10.1016/j.tra.2017.09.012
Fagnant D, Kockelman K (2015) Preparing a nation for autonomous vehicles: opportunities, barriers and policy recommendations. Transp Res Part A Policy Pract 77:167–181
Fagnant DJ, Kockelman K (2014) The travel and environmental implications of shared autonomous vehicles, using agent-based model scenarios. Transp Res Part C: Emerg Technol 40:1–13
Fraedrich E, Beiker S, Lenz B (2015) Transition pathways to fully automated driving and its implications for the sociotechnical system of automobility. Eur J Futur Res 3(1):11
Gettman D, Lott JS, Goodwin G, Harrington T (2017) Impacts of laws and regulations on CV and AV technology introduction in transit operations. Federal Highway Administration
Guerra E (2016) Planning for cars that drive themselves: metropolitan planning organizations, regional transportation plans, and autonomous vehicles. J Plan Educ Res 36(2):210–224
Guston DH (2014) Understanding anticipatory governance. Soc Stud Sci 44(2):218–242
Hajer M (2003) Policy without polity? Policy analysis and the institutional void. Policy Sci 36 (2):175–195
Harper CD, Hendrickson CT, Mangones S, Samaras C (2016) Estimating potential increases in travel with autonomous vehicles for the non-driving, elderly and people with travel-restrictive medical conditions. Transp Res Part C Emerg Technol 72:1–9
JafariNaimi N (2018) Our bodies in the trolley's path, or why self-driving cars must *not* be programmed to kill. Sci Technol Hum Values 43(2)
Jasanoff S (2003) Technologies of humility: citizen participation in governing science. Minerva 41 (3):223–244
Jasanoff S (2016) The ethics of invention: technology and the human future. WW Norton & Company
Jokinen JP, Sihvola T, Mladenovic MN (2017) Policy lessons from the flexible transport service pilot Kutsuplus in the Helsinki Capital Region. Transp Policy. https://doi.org/10.1016/j.tranpol.2017.12.004
König M, Neumayr L (2017) Users' resistance towards radical innovations: The case of the self-driving car. Transp Res Part F Traffic Psychol Behav 44:42–52
Latour B, Venn C (2002) Morality and technology. Theory Cult Soc 19(5–6):247–260
Lutin J, Kornhauser A, Lerner-Lam E (2013) The revolutionary development of self-driving vehicles and implications for the transportation engineering profession. ITE J 83(7):28–33
Lyons G, Davidson C (2016) Guidance for transport planning and policymaking in the face of an uncertain future. Transp Res Part A Policy Pract 88:104–116

Metzger J, Allmendinger P, Oosterlynck S (2015) Planning against the political: democratic deficits in European Territorial Governance. Routledge, London

Meyer J, Becker H, Bösch PM, Axhausen KW (2017) Autonomous vehicles: the next jump in accessibilities? Res Transp Econ 62:80–91

Milakis D et al (2017a) Development and transport implications of automated vehicles in the Netherlands: scenarios for 2030 and 2050. Eur J Transp Infrastruct Res 17(1):63–85

Milakis D, van Arem B, van Wee B (2017b) Policy and society related implications of automated driving: a review of literature and directions for future research. J Intell Transp Syst 21(4): 324–348

Mladenovic M, Blyth PL, Abbas M, Kosonen I (2016) Intersecting our mobilities: path dependence from manually-operated semaphore to self-driving vehicles? In: 2016 IEEE international symposium on technology and society (ISTAS). IEEE, New York, pp 1–6

Mladenovic M et al (2017) Is planning (for) self-driving vehicle technology a question of social justice? A Finnish perspective. Association of European Schools of Planning. Lisbon

Nieuwenhuijsen J, de Almeida Correia GH, Milakis D et al (2018) Towards a quantitative method to analyze the long-term innovation diffusion of automated vehicles technology using system dynamics. Transp Res Part C Emerg Technol 86:300–327

Owen R, Macnaghten P, Stilgoe J (2012) Responsible research and innovation: from science in society to science for society, with society. Sci Public Policy 39(6):751–760

Pacey A (1999) Meaning in technology. MIT Press, Cambridge

Pangbourne K, Stead D, Mladenovic M, Milakis D (2018) The case of mobility as a service: a critical reflection on challenges for urban transport and mobility governance. In: Marsden G, Reardon L (eds) Governance of the smart mobility transition. Emerald, pp. 33–50

Quay R (2010) Anticipatory governance. J Am Plan Assoc 76(4):496–511

Rayner S (2004) The novelty trap: why does institutional learning about new technologies seem so difficult? Ind High Edu 18(6):349–355

Rip A (1986) Controversies as informal technology assessment. Knowledge 8(2):349–371

Schellekens M (2015) Self-driving cars and the chilling effect of liability law. Comput Law Secur Rev 31(4):506–517

Schiller P (2016) Automated and connected vehicles: high tech hope or hype? World Transp Policy Pract 22(3):28–44

Shladover SE (2017) Connected and automated vehicle systems: introduction and overview. J Intell Transp Syst, 1–11

Stilgoe J (2017) Machine learning, social learning and the governance of self-driving cars. Soc Stud Sci, 30631271774168

Stilgoe J, Owen R, Macnaghten P (2013) Developing a framework for responsible innovation. Res Policy 42(9):1568–1580

Talebpour A, Mahmassani HS (2016) Influence of connected and autonomous vehicles on traffic flow stability and throughput. Transp Res Part C Emerg Technol 71:143–163

Thomopoulos N, Givoni M (2015) The autonomous car—a blessing or a curse for the future of low carbon mobility? An exploration of likely vs. desirable outcomes. Eur J Futur Res 3(1):14

Upham P, Kivimaa P, Virkamäki V (2013) Path dependence and technological expectations in transport policy: the case of Finland and the UK. J Transp Geogr 32:12–22

Urry J (2004) The system of automobility. Theory Cult Soc 21(4–5):25–39

Vellinga N (2017) From the testing to the deployment of self-driving cars: legal challenges to policymakers on the road ahead. Comput Law Secur Rev 33(6):847–863

Voss J-P, Bauknecht D, Kemp R (2006) Reflexive Governance for sustainable development. Edward Elgar

Walker W, Marchau V (2017) Dynamic adaptive policymaking for the sustainable city: the case of automated taxis. Int J Transp Sci Technol 6(1):1–12

Williams T, Wagner J, Morgan C et al (2017) Transportation planning implications of automated/ connected vehicles on Texas highways (0-6848-1). Texas Department of Transportation

Winner L (1977) Autonomous technology: technics-out-of-control as a theme in political thought. MIT Press, Cambridge

Zakharenko R (2016) Self-driving cars will change cities. Reg Sci Urban Econ 61:26–37
Zmud J, Goodin G, Moran M et al (2017) Advancing automated and connected vehicles: policy and planning strategies for state and local transportation agencies. Federal Highway Administration

Miloš N. Mladenović is an assistant professor at the Spatial Planning and Transportation Engineering Group, Department of Built Environment, Aalto University, Finland. He obtained his B.Sc. in Transport Engineering from the University of Belgrade, Serbia, and his M.Sc. and Ph.D. in Civil Engineering from Virginia Tech, USA. He has also held a visiting research position at the Faculty of Architecture and the Built Environment, Delft University of Technology, the Netherlands. His current research interests include ethical assessment of emerging mobility technologies, socially sustainable planning methods, asset management methods for intelligent transport systems, and engineering education assessment practices. His previous and current teaching responsibilities include a range of courses in transport systems planning and policy, transport modeling, traffic management, and geometric design. He is the organizing professor of the annual Aalto University Summer School on Transportation.

Part III
Governance of Electric Mobility

Chapter 7
A Comparison of Policy Measures Promoting Electric Vehicles in 20 Countries

Nele Rietmann and Theo Lieven

Abstract This chapter compares and analyzes policy measures that governments have implemented in 20 countries, on five continents, to promote consumer adoption of electric vehicles (EVs). These include subsidies for EV purchases, tax benefits, incentives for developing charging infrastructures, and traffic regulations such as free parking or the use of bus lanes. To assess the effectiveness of these measures, this study examines their influence on the market share of EVs in each country. Results indicate that the higher the level and amount of political incentives in a country, the higher the country's proportion of EVs. This is analyzed further in case studies of the Netherlands and Brazil by evaluating the countries' governance systems. While the Netherlands has benefited from a collaborative effort of public and private actors promoting EVs, Brazil still faces many challenges. Altogether, findings show that policy measures have a significant impact on EV adoption around the world and that stakeholders from different sectors must work together to maximize the effectiveness of these policies. Therefore, countries wishing to raise EV penetration should do so through a range of governmental incentives and a collaborative governance system.

Keywords Electric vehicles · Policy measures · International comparison Governance systems

7.1 Introduction

Governmental support for electric vehicles (EVs) has been on the rise in many countries over the past several years. Yet, apart from a few exceptions, such as Norway and the Netherlands, consumer sales of EVs are still rather low in most countries

N. Rietmann (✉) · T. Lieven
Institute for Customer Insight, University of St. Gallen, St. Gallen, Switzerland
e-mail: nele.rietmann@unisg.ch

T. Lieven
e-mail: theo.lieven@unisg.ch

M. Finger and M. Audouin (eds.), *The Governance of Smart Transportation Systems*, The Urban Book Series, https://doi.org/10.1007/978-3-319-96526-0_7

around the world. The majority of market shares still lie significantly below 2% (OECD/IEA 2016). Differences exist in the forms and levels of national governments' consumer incentives and market shares of EVs (EAFO 2017a; EV-Volumes 2017; IEA IA-HEV 2016; OICA 2017a). Using a cross-national perspective to compare these diverse policy measures is valuable to analyze their effectiveness. It is also important to note that there are different levels at which EV penetration is promoted. For example, there are international agreements, such as initiatives by the International Energy Agency (IEA) or emission standards set by the European Union (EU). National measures include financial subsidies, tax incentives, or legislation regarding EVs. Furthermore, there are measures at regional and local levels, such as the largest Dutch cities offering monetary incentives on top of the national ones (van der Steen et al. 2015). Nonetheless, the focus in this analysis is at the country level, since most policies are set by national governments and this perspective allows for comparing policy strategies in different countries. Therefore, it is possible to draw generalizable conclusions regarding effective policy measures supporting EV penetration than would be possible with a focus on more global or local cases.

Previous research has predominantly focused on policy measures in cities, a single country, or a comparison between two countries (e.g., Zhang et al. 2011; Bakker and Trip 2013; Holtsmark and Skonhoft 2014; Figenbaum et al. 2015; Li 2016). For example, Bakker and Trip (2013) investigate how cities can effectively promote EVs with policy measures. They find that municipal governments can support the adoption process of EVs through policies such as investments in charging infrastructure. However, their findings also show that most lobbying in this regard occurs at the regional and national levels, which further highlights the importance of focusing on a country level. Moreover, Figenbaum et al. (2015) take a closer look at the successful adoption of EVs in Norway and investigate how policy measures contribute to this. Results demonstrate that various different factors influenced the successful diffusion of EVs—most importantly the large incentives offered by the Norwegian government. Li (2016) also highlights the importance of governmental policies supporting EVs, particularly regarding infrastructure developments, in a comparative study examining China and Brazil. The author points out that China is far ahead of Brazil in terms of consumer adoption of EVs, due to stronger policies and infrastructure investments.

In addition to this, some studies have also taken international perspectives to investigate policy measures promoting EV penetration (e.g., Sierzchula et al. 2014; Haddadian et al. 2015; Lieven 2015). Sierzchula et al. (2014) were among the first to examine the influence of policy measures on EV adoption across a range of countries. Using data from 2012, the authors analyzed the impact of financial incentives on the market share of EVs in 30 countries. While their findings revealed a positive correlation of monetary incentives and charging infrastructure with EV market shares, they found no support for a causal effect. One of the main reasons for this is certainly that market shares were still very low in 2012, with EVs accounting for only approximately 0.35% in all 30 countries that were included in this analysis. This has changed significantly over the past few years. Norway, which had an EV market share of approximately 3.1% in 2012, reached a market share of nearly 30%

in 2016 (EV-Volumes 2017; OICA 2017a; Sierzchula et al. 2014). It is important to investigate how policies have recently influenced this sales growth. Another global study (Lieven 2015) examined the effects of policy measures promoting EVs by looking at consumer preferences in 20 countries on five continents. Results indicate that consumers appreciate financial incentives but consider a sufficient charging network on freeways to be an absolute must-have. Other policy measures, such as free access to fast lanes or bus lanes, received less positive responses from consumers. Overall, findings suggest that policies supporting the development of a charging infrastructure as well as monetary incentives should be the most effective incentives to promote EV adoption.

Building on Lieven's (2015) findings, this chapter will take a closer look at policy measures promoting EV penetration in the same 20 countries considered in the previous research. It will do so by first comparing the different incentives that governments implemented in these countries and then investigating their influence on EV market shares. Based on this analysis, two cases will be explored more closely—the Netherlands and Brazil—to determine in more detail how national governance systems have supported or inhibited EV penetration in these countries. This chapter aims to provide an international overview of the different types and levels of incentives that governments offer to promote EV penetration in 20 countries. It also discusses how to assess the successfulness of these measures. In order to do so, the following hypotheses, formulated on the basis of prior research findings, will be tested:

H1: Monetary measures, traffic regulations, and infrastructure measures to promote EVs have a positive impact on the market share of EVs.
H2: The interaction of monetary measures and a country's charging infrastructure has the strongest positive influence on the market share of EVs.

7.2 Policy Measures Promoting EVs in 20 Countries

These hypotheses will be tested in the following sections to determine the impact of various policy measures on EV penetration in 20 countries. First, it will be outlined how data was collected and how certain figures were determined for this analysis. An overview of the different types of policy measures that have been implemented in the 20 countries will then be presented. Finally, a structural equation model will allow for testing the previously formulated hypotheses.

7.2.1 Data Collection

In order to compare policy measures across nations and construct a structural equation model assessing their influence on EV market shares, data was gathered from a range of secondary sources for 20 countries (Australia, Belgium, Brazil, Canada, China, France, Germany, Hong Kong, India, Italy, Japan, Korea, the Netherlands, Norway, Russia, South Africa, Switzerland, Taiwan, the UK, and the USA). Data was collected for both battery electric vehicles (BEVs) and plug-in hybrid electric vehicles (PHEVs), as this dataset was assumed to represent all consumers interested in cars with electric propulsion. In this chapter, EVs refer to both BEVs and PHEVs.

With regard to policy measures, all forms of incentives that directly affected consumers purchasing EVs were considered. This included financial measures, such as purchase subsidies, tax benefits, and other financial benefits (such as import-duty exemptions), infrastructure incentives (such as tax deductions for installing charging stations), and traffic regulations (such as the free use of fast or bus lanes or free parking). By focusing on consumer incentives, as opposed to incentives for companies to adopt EVs, it is possible to evaluate the impact of these measures on EV market shares from a consumer perspective.

The figures for EV market shares were calculated from the total number of EVs sold in each of the 20 countries in 2016. This data was provided by a Swedish consulting firm (EV-Volumes 2017) as a percentage of the total amount of vehicles sold during this period, according to the Organisation Internationale des Constructeurs d'Automobiles (OICA 2017a). The OICA differentiates between commercial vehicles and passenger cars in its data. Since this study focuses on incentives aimed at consumers and their adoption of EVs, only the number of passenger cars was included in the market share calculations.

The subsequent analysis of the fiscal measures was complex, since there are significant differences between countries and regional divergences within nations. Some countries offer a onetime tax benefit for EV purchasers, while others provide annual tax benefits for EV owners. For example, the Norwegian government allows a onetime exemption from the purchase tax for BEVs and a reduction for PHEVs (EAFO 2017a). In Switzerland, there are lowered, annual road taxes for electric cars (Bundesamt für Energie 2017). To compare countries with such different policy measures, the figures indicate the sum of all fiscal measures in a given country for a number of years. This assumes ownership for approximately 10 years and that currencies, such as the Euro or the American dollar, remain approximately equal during the time period. The figures in this analysis should primarily be interpreted as indicators for a country's incentive level.

In addition, non-monetary policies, specifically traffic regulations such as free access to fast or bus lanes, and infrastructure investments, were expressed in terms of dummy variables in the structural equation model analyzing the influence of policy measures on EV market shares. The respective variables show the existence, or nonexistence, of non-monetary measures and the relative extent to which they are present in a particular country.

The number of charging stations was collected for each country, in order to analyze the extent to which such incentives promoted the development of national charging networks. Various sources were compared for this (ChargeHub 2017; EAFO 2017a, b; Environmental Protection Department 2017; OECD/IEA 2016;

PlugShare 2017), since it was not always possible to determine from just one source if figures indicated single charger or station with one or more charging spots. This comparison indicates the pervasiveness of charging networks in the 20 countries. Furthermore, the number of charging stations was expressed in relative terms to compare the different nations. Specifically, a country's total number of charging stations was divided by its population (Worldometers 2017) and by its length of highway kilometers (NationMaster 2017). These two figures were both taken into account in the structural equation model for the variable indicating the number of charging stations.

7.2.2 *Findings*

Based on the data gathered on existing policy measures promoting EV penetration in the 20 countries, an overview was constructed to show the types of incentives that exist in each country, as Table 7.1 shows. All countries, except Australia and

Table 7.1 Overview of national policy measures promoting EV penetration in 20 countries in 2016

	Purchase subsidies	Tax benefits	Other financial benefits	Infrastructure measures	Traffic regulations
Australia					
Belgium	x	x		x	
Brazil		x	x		
Canada	x			x	x
China	x	x			x
France	x	x		x	
Germany	x	x			x
Hong Kong		x			
India	x	x	x		
Italy	x	x		x	x
Japan	x				
Korea	x	x			x
Netherlands		x		x	
Norway		x		x	x
Russia			x		
South Africa					
Switzerland		x			
Taiwan		x			
UK	x	x		x	x
USA	x	x		x	x

South Africa, offered some form of incentive for consumers to purchase EVs in 2016. These two exceptions were taken into account for the subsequent analysis investigating the influence of policy measures on market shares. By including countries that do not offer incentives, the data can show a more complete picture. The different types of political incentives, specifically monetary measures, infrastructure measures, and traffic regulations, will first be outlined in more detail in a cross-country comparison.

7.2.2.1 Monetary Measures

Monetary measures include incentives such as purchase subsidies and tax benefits for EV buyers. As can be seen from Table 7.1, these are the most common measures taken by governments in the 20 countries considered in this analysis. In fact, all of the countries, except Australia and South Africa, offer some form of fiscal incentive for EV buyers. Nonetheless, there are significant cross-country differences between these fiscal measures. For example, Belgium and Canada both offer financial incentives but primarily on a regional basis. In the USA, there are also differences between states. In addition to the national tax incentive of US$2500 to US$7500 (Office of Energy Efficiency & Renewable Energy 2018), a number of states have also implemented statewide subsidies. For example, California offers additional rebates of up to US $2500 (van der Steen et al. 2015). However, most other countries predominantly offer incentives at only a national level. For example, Germany, China, France, India, Italy, Korea, and the UK have implemented a combination of both purchase subsidies and tax benefits. The values and extents of these benefits differ significantly (Beltramello 2012; EAFO 2017a; Government of NCT of Delhi 2016; IEA IA-HEV 2014, 2016; Kim and Yang 2016; OECD/IEA 2016; Rokadiya and Bandivadekar 2016; van der Steen et al. 2015). In addition, various changes have already occurred in terms of the types and levels of incentives offered over time. For example, while Japan used to have a combined policy of subsidies and tax benefits in the form of acquisition and tonnage tax exemptions between 2009 and 2012 (JAMA 2010a), it now only offers purchase subsidies under the Green Vehicle Purchasing Promotion and the Clean Energy Subsidy program (Hao et al. 2014; JAMA 2010b). Therefore, policy strategies should not be considered fixed, since governments adapt them over time.

Moreover, countries such as Brazil, Hong Kong, the Netherlands, Norway, Switzerland, and Taiwan do not offer any national subsidies but different forms of tax benefits. For example, EV owners in Brazil have benefited from a reduction in the Motor Vehicle Property Tax (MVPT) since 2014, the level of which differs between regions within the country (such as a 50% reduction in São Paulo) (Benvenutti et al. 2016; Prefeitura de São Paulo 2015). In the Netherlands, EV owners are exempt from registration and road taxes (50% reduction for PHEVs), although a slow phaseout of fiscal incentives for PHEVs has been announced for the period between 2017 and 2020 (EAFO 2017a; IEA IA-HEV 2013). Other countries offer purchase, registration, and/or annual road-tax exemptions or reductions. In Norway, this greatly affects the price of EVs, since the purchase tax, which is the same as the country's import tax, equals 25% (EAFO 2017a). Combined with an annual road tax that is reduced to 455 kroner (approximately US$55) from 2820 or 3290 kroner (approximately US$344 or US$400) for conventional vehicles (The Norwegian Tax Administration 2017). This can add up to a total fiscal incentive of approximately US$25,000 over 10 years.

In addition, some countries have implemented a number of other financial incentives. For example, Brazilian EV purchasers have been exempt from the import duty since 2015 (Benvenutti et al. 2016), and the Russian government has allowed zero tariff on imported EVs since 2014 (Gazeta 2016).

Overall, there are significant differences in the types and levels of monetary measures that governments implemented to promote EV penetration in the 20 studied countries. Table 7.2 shows a more detailed overview of the total values of fiscal incentives.

7.2.2.2 Infrastructure Measures

In addition to these financial measures, various governments also provide incentives for developing a charging infrastructure. These incentives come in very different forms and, similar to the monetary measures, may also vary between regions in the same country. For example, Canadian EV owners in British Columbia may receive a reduction of up to CAN$500 for a residential electric charging station (IEA IA-HEV 2012). In Ontario, there is a 50% purchase and installation incentive of up to CAN $1000 (IEA IA-HEV 2013). In Québec, individuals may receive up to CAN$270 for the equipment for charging stations and CAN$195 for installation (IEA IA-HEV 2016). In the Netherlands, there are also some regional differences between Friesland, which offers a €500 discount on private charging points, and partial subsidies for charging spots in Rotterdam (van der Steen et al. 2015). Similarly, local governments and states in the USA (such as California) provide rebates and tax credits for charging stations (Berman 2017; van der Steen et al. 2015). Other countries, including Belgium, France, Italy, Norway, and the UK, offer nationwide incentives such as tax deductions or subsidies for private charging point installations, public funding, or the free use of charging infrastructure (EAFO 2017a; van der Steen et al. 2015). The relative extent to which countries offer such infrastructure measures is summarized in Table 7.2.

7.2.2.3 Traffic Regulations

Traffic regulations are often altered in favor of EV drivers to support the widespread sales of EVs. These regulations include access to high-occupancy-vehicle (HOV) lanes, highways, toll roads, bridges, and ferries in places such as Canada, the USA, and Norway (EAFO 2017a; IEA IA-HEV 2008, 2016). Moreover, EVs are offered free, or reduced-fee, parking in urban areas in Germany, Italy, Korea, Norway, some local areas in the UK, and in certain states in the USA (Berman 2017; EAFO 2017a; IEA IA-HEV 2011, 2016; Kim and Yang 2016). Other

Table 7.2 Data for 20 countries

Country	EVs sold in 2016	Total sales 2016	EV market share 2016 (%)	Monetary measures	Infrastructure measures	Traffic regulation	Charging stations total	Pop. in millions	Chargers per 1 mio. population	Highway km	Chargers per highway km
Australia	1433	927,274	0.15	0	0	0	60	24,642	2.43	1700	0.04
Belgium	9598	539,519	1.78	6500	1	0	1715	11,444	149.86	1763	0.97
Brazil	48	1,676,722	0.00	500	0	0	60	211,243	0.28	11,000	0.01
Canada	10,842	661,088	1.64	6000	1	1	4465	36,626	121.91	16,900	0.26
China	350,578	24,376,902	1.44	6000	0	2	81,000	1,388,233	58.35	97,355	0.83
France	34,205	2,015,177	1.70	7000	1	0	16,000	64,939	246.39	11,392	1.40
Germany	28,079	3,351,607	0.84	4000	0	0.2	24,667	80,636	305.91	12,845	1.92
Hong Kong	1367	41,600	3.29	8000	0	0	1500	7402	202.65	100	15.00
India	1023	2,966,637	0.03	3000	0	0	328	1,342,513	0.24	1208	0.27
Italy	3463	1,824,968	0.19	5000	1	0.5	2205	59,798	36.87	6661	0.33
Japan	23,438	4,146,459	0.57	5000	0	0	40,000	126,045	317.35	7383	5.42
Korea	6108	1,533,813	0.40	5000	0	0.5	1270	50,705	25.05	4044	0.31
Netherlands	23,129	382,825	6.04	20,000	1	0	28,246	17,033	1658.31	2808	10.06
Norway	45,685	154,603	29.55	25,000	2	3	8734	5331	1638.40	664	13.15
Russia	416	1,239,680	0.03	2000	0	0	200	143,375	1.39	1400	0.14
S. Africa	159	361,289	0.04	0	0	0	40	55,436	0.72	1927	0.02
Switzerland	5654	317,318	1.78	5000	0	0	3828	8454	452.80	1361	2.81

(continued)

Table 7.2 (continued)

Country	EVs sold in 2016	Total sales 2016	EV market share 2016 (%)	Monetary measures	Infrastructure measures	Traffic regulation	Charging stations total	Pop. in millions	Chargers per 1 mio. population	Highway km	Chargers per highway km
Taiwan	2	206,092	0.00	2000	0	0	360	23,405	15.38	1335	0.27
UK	39,351	2,692,786	1.46	6000	1	1	12,320	65,511	188.06	6016	2.05
USA	158,212	6,872,729	2.30	7500	2	1	42,918	326,474	131.46	75,008	0.57

Sources EV sales (EV-Volumes 2017); total sales (OICA 2017a); monetary measures (Beltramello 2012; Benvenutti et al. 2016; Bundesamt für Energie 2017; EAFO 2017a; Environmental Protection Department 2017; Gazeta 2016; Government of NCT of Delhi 2016; Hao et al. 2014; IEA IA-HEV 2008, 2012, 2013, 2014, 2016; JAMA 2010b; Kim and Yang 2016; Maurer 2014; Mock and Yang 2014; OECD/IEA 2016; Prefeitura de São Paulo 2015; PwC 2016; Rokadiya and Bandivadekar 2016; van der Steen et al. 2015); infrastructure measures (Berman 2017; EAFO 2017a; IEA IA-HEV 2012, 2013, 2016; van der Steen et al. 2015); traffic regulations (Berman 2017; EAFO 2017a; IEA IA-HEV 2008, 2011, 2016; Kim and Yang 2016; OECD/IEA 2016); charging stations (ChargeHub 2017; EAFO 2017a, b; Environmental Protection Department 2017; OECD/IEA 2016; PlugShare 2017); population (Worldometers 2017); highway km (NationMaster 2017) incentives include free use of bus lanes, such as in Germany and Norway (EAFO 2017a). China has implemented a policy of free, unrestricted number plates for EVs in cities such as Beijing or Shanghai, where it is otherwise very difficult to receive a number plate with a conventional vehicle (OECD/IEA 2016; Zhu et al. 2017). The UK introduced another urban policy in 2013, in which EVs are exempt from the congestion charge in London, which can otherwise cost up to an annual £2000 (EAFO 2017a). Further incentives are also offered at local levels, such as in Korea, the UK, and the USA (Berman 2017; EAFO 2017a; Kim and Yang 2016). Table 7.2 shows the relative levels at which such policies have been implemented in the 20 countries.

7.2.2.4 Influence of Policy Measures on EV Market Shares

In order to determine the effectiveness of policy measures, their influence on EV market shares was calculated. This was done by first adding together all monetary measures for each country, and expressing infrastructure measures and traffic regulations in terms of dummy variables, indicating the existence and relative strength of these measures in a given country. Moreover, the number of chargers, as an indicator for the prevalence of a charging infrastructure, was included in the analysis. Table 7.2 summarizes all of the data. The variables in bold were used for further analysis. Specifically, a structural equation model with partial least squares regression (PLS) was constructed (Fig. 7.1). Due to the relatively small sample size

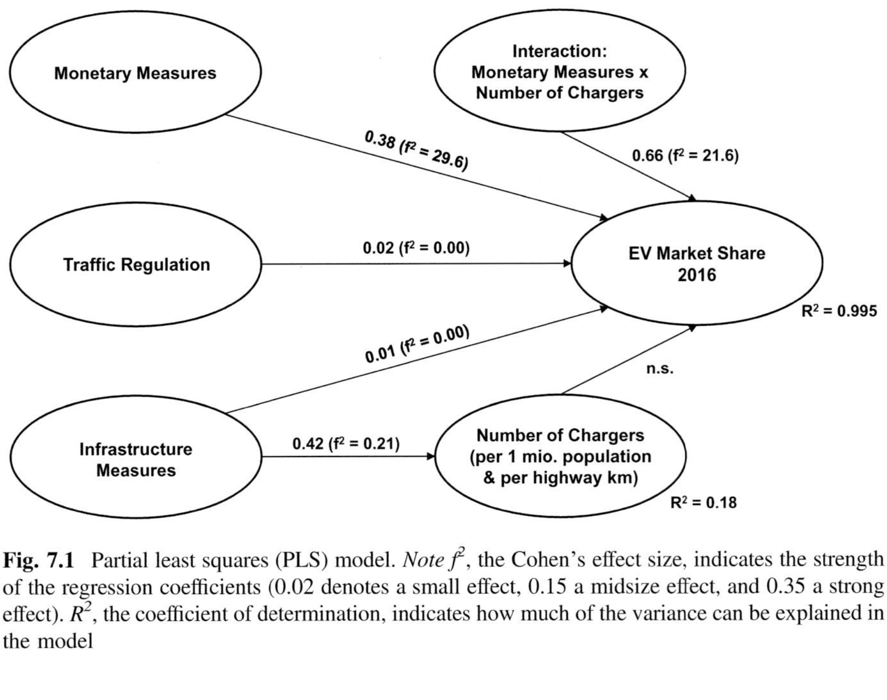

Fig. 7.1 Partial least squares (PLS) model. *Note* f^2, the Cohen's effect size, indicates the strength of the regression coefficients (0.02 denotes a small effect, 0.15 a midsize effect, and 0.35 a strong effect). R^2, the coefficient of determination, indicates how much of the variance can be explained in the model

of 20 countries, this analysis reports Cohen's effect size f^2 (Cohen 1988). This measure is independent of sample size (Selya et al. 2012) and represents the change in R^2, the coefficient of determination, when a given effect is included in the model.

In the base model with monetary measures, traffic regulations, infrastructure measures, and existing charging stations, both monetary incentives and regulation measures have a positive effect on EV market share ($\beta_{\text{Monetary Measures}} = 0.60$; $f^2 = 0.34$; $\beta_{\text{Traffic Regulation}} = 0.21$; $f^2 = 0.11$). Therefore, the higher the incentives that the governments provide, the higher the market share of EVs in the respective countries. Moreover, the effect of infrastructure measures on the EV market share appears to be mediated by the number of charging stations ($\beta_{\text{Chargers}} = 0.42$; $f^2 = 0.21$), which have a positive effect on EV market share ($\beta_{\text{Infrastructure-Chargers}} = 0.22$; $f^2 = 0.09$). Therefore, the higher the incentives for charging infrastructure are, the higher the number of chargers. Altogether, this results in a higher EV market share. Since all three measures show a positive impact on the percentage of EV registrations in 2016, H1 is supported.

Taking monetary measures, traffic regulations, and infrastructure measures with the number of chargers, approximately 89% of the variance in the market share of EVs can be explained by the model ($R^2 = 0.887$). Figure 7.1 depicts a further analysis, which includes an interaction term of the monetary measures and the number of chargers. This interaction effect has a positive influence on the market share of EVs ($\beta_{\text{Interaction}} = 0.66$; $f^2 = 21.6$). The effect is very strong, even to the extent that it suppresses the impacts of traffic regulations, infrastructure, and the direct effect of the number of charging stations. Overall, the effect raises the explained variance in the market share of EVs to almost 100% ($R^2 = 0.995$).

Therefore, the analysis with PLS shows that all policy measures, particularly monetary incentives, have a positive effect on EV market shares. Moreover, monetary measures, in combination with a sufficient charging infrastructure, have an even stronger overall impact on the percentage of EVs in a country. Therefore, H2 is also supported.

As the comparison and analysis of policy measures promoting EV penetration in 20 countries have shown, significant differences exist between countries in terms of the types and levels of available incentives and consumer sales of EVs. Based on this analysis, it is recommended that governments which want to support a more widespread adoption of EVs should do so through a strategy of combined policy measures. In particular, monetary incentives in the form of subsidies or tax benefits may provide an initial stimulus for consumers to consider purchasing an EV rather than a conventional vehicle. However, as Lieven's (2015) consumer study showed, a sufficient charging infrastructure is also a key. Governments should not only focus on short-term incentives but also take a more long-term perspective and provide investments for necessary EV charging networks.

7.3 More Successful and Less Successful Governance Systems for EVs

As the previous analysis shows, policy measures have a significant influence on EV market shares. The data in Table 7.2 also highlights the strong differences between countries, in terms of EV incentives and market shares. In particular, the Netherlands (with the second-highest EV market share) and Brazil (with the second-lowest EV market share) showed some considerable differences in the types and levels of incentives that the respective governments implemented. In fact, the Dutch government provides approximately 40 times higher a monetary incentive than does Brazil. In addition, the Netherlands also implemented a number of infrastructure measures, which was not the case in Brazil.

Apart from the policy measures discussed above, there are certainly other factors that influenced the stronger penetration of EVs in the Netherlands and the lower market shares in Brazil. The following case studies will discuss how the governance systems in these two countries supported or hindered a strong EV penetration. In particular, this chapter will consider which actors were involved in this process, and which actions might have contributed to the relative successfulness or unsuccessfulness of EV sales among consumers in the Netherlands and Brazil. The following case studies will show that the two countries offer a valuable comparison, as the Netherlands has taken a variety of proactive measures at the national and local levels to support EV penetration. These have been supported by various stakeholders, which strengthened their effectiveness. Brazil, on the other hand, lacks the necessary support at the country and supranational levels to effectively promote EVs among consumers.

7.3.1 Case Study: The Netherlands

The Netherlands has seen significant growth in its number of EVs over the past several years and become a global leader in EV sales (IEA IA-HEV 2016). In fact, it has the second-highest EV market share worldwide, after Norway (Mock and Yang 2014). Having set a target for 2015 of 15,000 to 20,000 EVs on Dutch roads, this was largely exceeded, with a total of 90,000 EVs by the end of 2015 (IEA IA-HEV 2016). A key reason for this is the high monetary incentive that the government has implemented, as van't Hull and Linnenkamp (2015) highlighted. Results from the previous analysis further support this. According to van't Hull and Linnenkamp (2015), various projects have also been implemented by different cities and regions in the Netherlands, such as offering additional subsidies to meet European air quality targets for 2015. More recently, the 2016–2020 Electric Transport Green Deal was signed by 18 parties, including the central Dutch government, the Formula E Team, and a variety of other market players (Netherlands Enterprise Agency 2017). A central aim of the Green Deal is that 10% of new passenger cars must have electric power. In addition to various policies and initiatives promoting EVs in the Netherlands, another important driver has been collaboration between various actors, including national, regional, and local governments, and different companies, organizations, and other stakeholders. This will be highlighted further in this case study by first analyzing additional initiatives on top of national policies, and the development of the charging infrastructure in the Netherlands. This case study will then provide an overview of the actors involved in promoting EVs to consumers.

First, it is important to note that there are other policy measures and initiatives in the Netherlands in addition to the national ones. For example, Amsterdam and Utrecht provide fiscal incentives on top of the ones offered by the central government (van der Steen et al. 2015). While this was taken into account to some extent in the previous analysis, local measures provide additional awareness of EV incentives and, most likely, greater willingness for potential buyers to purchase an EV. Further initiatives have started at the local level, such as the MRA-Electric (Amsterdam Metropolitan Area Electric, MRA-E), initiated by local Amsterdam authorities to support the widespread rollout of electro-mobility in the Amsterdam metropolitan area (MRA) (van't Hull and Linnenkamp 2015). In collaboration with the government and the market, this project contributed to developing a charging infrastructure in the area. Moreover, carsharing systems with EVs have significantly expanded over the past several years (IEA IA-HEV 2016). For example, Daimler has set up its car2go system in Amsterdam, with 300 electric Smart Fortwo cars (van't Hull and Linnenkamp 2015). While this does not directly contribute to passenger car sales, it does bring consumers in closer contact with EVs by providing opportunities for firsthand experiences with such vehicles. Initiatives such as this may also indirectly contribute to greater EV purchase. Similarly, introducing large numbers of EVs in airport taxi fleets at Schiphol (van't Hull and Linnenkamp 2015) may have further contributed to growing EV acceptance in the Netherlands. Finally, the Vereniging Elektrische Rijders (VER) initiative is an association for EV drivers, which acts as an information and linking platform for continuous support of electric cars (IEA IA-HEV 2016). Therefore, EV owners also receive non-governmental support and necessary information to ease the adoption process.

In addition, there have been significant improvements in developing charging infrastructure in the Netherlands, which has certainly also contributed to growing EV market shares. For example, the Green Deal, which was signed in 2015, provides governmental support of €5.7 million for developing a national charging infrastructure (IEA IA-HEV 2016). However, approximately 80% of the charging stations were installed beforehand by the E-Laad Foundation, a partnership among Dutch grid operators that provided municipalities with free chargers (Bakker and Trip 2015). All other chargers were primarily installed by larger cities such as Amsterdam, Rotterdam, Utrecht, and The Hague (Bakker and Trip 2015; van't Hull and Linnenkamp 2015). There are also further initiatives by the private sector to support a national charging network. For example, the German retailer Lidl set up fast-charging stations for EVs at a number of its stores, in collaboration with ABB (Netherlands Enterprise Agency 2017). The charging infrastructure in the

Netherlands has expanded over the past several years through initiatives and collaborations of various actors in the public and private sectors.

Indeed, one key factor contributing to growing EV sales in the Netherlands has been the collaboration of many actors that have supported national EV penetration. For example, more than 40 parties signed the National Energy Agreement for Sustainable Growth, which aims to decrease CO_2 emissions in the mobility sector by 17% by 2030, and by 60% by 2050 (EAFO 2017a). Bakker et al. (2014) highlighted the importance of collaborations among various actors to achieve a more widespread adoption of EVs in their study on stakeholder interests, expectations, and strategies in the Netherlands to promote EV penetration. The authors conducted interviews with a variety of stakeholder representatives, including central and local governments, automotive manufacturers and importers, electricity producers, and various others. Findings showed that there is general support by these stakeholders to develop an EV system, even though their individual agendas differ. This is due to strong regulatory pressure on the automotive industry, pushing manufacturers to construct EVs. The various stakeholders took this as an opportunity to engage in EVs from early on and continuously learn from its impacts. Bakker et al. (2014) further note important factors that have contributed to the development of EVs, specifically in the Dutch context. A key reason why EVs have received support from various actors in the Netherlands is that the country does not have major car manufacturers, allowing the government to act more independently than is possible in other countries in which industrial interests are much more prevalent and inhibiting regarding consumer purchase of EVs. Overall, this case shows that the successful adoption of EVs in the Netherlands has been driven by political incentives, non-policy actions, and the country's wider governance system, in which public and private agents collaborate to seize opportunities for developing a national EV system.

7.3.2 Case Study: Brazil

Compared to the Netherlands, the EV system in Brazil is much less developed. EVs make up an insignificant portion of the country's total car fleet to date (Benvenutti et al. 2016). Although there has been some growth over the past several years, Brazil still lags behind other countries, such as China and most developed countries in the world (Li 2016). Latin America generally faces a number of barriers regarding growth in EV penetration, given the lack of incentives, the high cost of EVs, insufficient charging infrastructure, and grid reliability issues (Marchán and Viscidi 2015; Teixeira et al. 2015). Nevertheless, Brazil is a particularly interesting case to investigate, since it offers various possibilities that could be beneficial for EVs. For example, more than 85% of Brazil's electricity is produced from renewable sources, and the country has further committed to reduce greenhouse gas (GHG) emissions in the future (Marchán and Viscidi 2015). Nonetheless, various factors, particularly institutional barriers (Domingues and Pecorelli-Peres 2013), have inhibited EV sales in Brazil.

As the prior analysis shows, no direct subsidy is offered to Brazilian consumers for EVs (Benvenutti et al. 2016). Furthermore, current incentives are not sufficient for consumers to purchase an EV, due to the many perceived disadvantages of EVs (Marchán and Viscidi 2015). Although federal tax incentives may have slightly stimulated EV adoption, various obstacles still exist.

The high price of EVs in Brazil is one of these key issues inhibiting the more widespread adoption of EVs by consumers. Although tax reductions are in place, they still account for a significant proportion of the purchase price (Baran and Legey 2013; Marchán and Viscidi 2015). For example, a Renault Zoé costs approximately four times more than in France, since the Brazilian government offers no subsidy for EVs and taxes remain high, despite current incentives (Li 2016). Taking into consideration that Brazil's GDP per capita is less than a fourth of that in France (The World Bank 2017), the purchase price of EVs poses a significant, comparatively high barrier in Brazil.

Brazil's insufficient charging infrastructure is another major obstacle to diffusion of EVs. This is particularly an issue since Brazil is geographically large. Cars are often needed to go long distances, which necessitates a range that many EVs do not yet offer (Teixeira et al. 2015). While other countries, especially in Europe, have started to recognize the need to build a charging infrastructure as a means to raise EV penetration, the Brazilian government offers no support for developing such an infrastructure. As Li (2016) noted, this implies that investments in Brazil's charging network currently depend on the private sector. While companies such as CPFL Energia have started setting up charging points for EVs, particularly in the São Paulo area, much improvement is still required for consumers to be willing to pay the high EV purchase price (Li 2016). Greater collaboration is necessary between various actors, including energy companies, city planners, the government, and others, to develop a sufficient charging infrastructure in Brazil.

Moreover, various actors and economic factors still hinder growth of Brazilian EV penetration. In particular, there are strong lobbying groups with an interest in hampering the introduction of EVs in the country. Specifically, ethanol lobbying groups strongly oppose EVs due to concerns that it will lower ethanol consumption (Baran and Legey 2013; Marchán and Viscidi 2015). As a result, the government also has very little interest in promoting EV penetration (Li 2016). This lack of support is further reinforced by the country's current economic situation. Due to Brazil's recession that has been ongoing since 2015, the government does not consider funding for EVs and charging infrastructure to be a priority (Li 2016).

Furthermore, unlike the Netherlands, Brazil is one of the largest automobile producers worldwide (OICA 2017b). Although the Brazilian automobile sector has recently been negatively affected by the economic crisis, production significantly grew, from the 1970s until 2013 to 2014, prior to the downturn. This led to a rise in liquid fuel consumption, from which the Brazilian economy has also greatly benefited (Baran and Legey 2013). Despite the country's plans to reduce GHG emissions, which can be achieved through a stronger EV penetration, the national interest in EV technology investment is still low because of this (Domingues and Pecorelli-Peres 2013). Therefore, the dominance of the automobile and oil industries in Brazil steers governmental interests to some extent and contributes to the country's slow EV adoption process among consumers.

In contrast to the Netherlands, Brazil faces obstacles caused by various stakeholder interests, and political and economic challenges, all of which hinder more widespread consumer adoption of EVs. While some incentives were put in place by the government, more collaboration between different interest groups is required for consumers and stakeholders to fully understand the potential, economic, and environmental benefits of EVs and support stronger penetration (Marchán and Viscidi 2015). As Domingues and Pecorelli-Peres (2013) state, the government's public policies are very important, especially when combined with a collaborative effort from diverse sectors. Compared to the Netherlands, Brazil certainly faces a number of additional challenges due to its lower level of economic development, and its political and current economic situation. In addition, the level of support for EV penetration at the supranational level also creates quite different political landscapes for the Netherlands in the EU and Brazil in South America. This could cause the Netherlands to be more proactive than Brazil in terms of policy measures. For example, the EU has set mandatory targets to reduce carbon dioxide (CO_2) emissions from passenger cars. Specifically, EU member countries must ensure that the average emissions level of new cars is 95 grams of CO_2 per kilometer by 2021 (European Commission 2018). This puts a certain amount of pressure on EU countries to proactively implement environmentally beneficial policies. On the other hand, Brazil does not face such supranational pressure, which may further explain the country's lack of support for EV penetration. Overall, the comparison between the Netherlands and Brazil highlights the importance of having a strong governance system in place that facilitates the collaboration between a country's various interest groups to raise EV market shares.

7.4 Implications and Conclusion

The comparison of policy measures supporting the penetration of EVs in 20 countries, on five continents, and the subsequent analysis of their impact on EV market shares highlights the importance of governmental incentives in promoting EVs. There are significant differences in the types and levels of incentives that different countries have implemented, as the study's comparative analysis demonstrates. Nonetheless, it can be concluded that monetary measures, infrastructure incentives, and traffic regulations all have a positive influence on EV market shares. In particular, monetary measures and strong support for developing a charging infrastructure have been proven effective. These results also support Lieven's (2015) finding that consumers appreciate financial incentives, but that sufficient charging networks are must-haves for individuals to be willing to purchase an EV.

Moreover, an analysis of successful and less successful governance systems for EVs, based on the case studies of the Netherlands and Brazil, further emphasizes the importance of having a collaborative system in place that supports EV penetration. Although governmental incentives can be very effective, substantial obstacles may arise if other key stakeholders do not contribute to, or strongly oppose, the adoption of EVs. For example, one of the key issues inhibiting widespread EV sales in Brazil is the dominance of the automotive industry and lobbying groups, which use their power to hinder EV implementation.

This implies that, while policy measures such as monetary incentives and infrastructure investments are important in promoting EVs, governments wishing to raise the EV market should also create a collaborative system and take further initiatives, such as establishing informative platforms about EVs or starting charging infrastructure projects. The Brazilian government should attempt to work with the automotive industry and lobbying groups that are opposed to EVs, rather than formulating policies that support these groups' interests. This requires a long-term perspective and strategic outlook, especially in times of economic downturns. Moreover, it is essential for key stakeholders, such as national and local authorities, automobile producers, importers, electricity producers, grid operators, and various others, to set common goals and work together to achieve them.

In addition to implementing measures such as financial incentives to strengthen EV market shares, governments may also find it effective to take general measures in support of electric mobility, as high EV market shares do not necessarily imply that electric mobility is also high. In fact, EV penetration merely indicates how many EVs are sold, but not how many people actually use electric cars. Governments may take further measures to raise their country's electric mobility. This includes, for instance, supporting carsharing projects with EVs, similar to Daimler's car2go system in Amsterdam (van't Hull and Linnenkamp 2015). Aside from promoting electric mobility more generally, this may also allow consumers who are not yet willing to purchase an EV to gain firsthand experiences and overcome potential barriers such as range anxiety. Ultimately, such projects may also lead to greater EV penetration over time.

Overall, this research provides important implications for both the public and private sectors. Nonetheless, there are also a number of limitations that should be taken into account for future research in this area. For example, the small sample size of 20 countries can be seen as problematic. Given the large quantity of data that was gathered for each country, the number of cases was nonetheless considered sufficient here. In addition, the potential sample-size issue was taken into account by calculating Cohen's effect size, which was independent of sample size, instead of t values in the structural equation model.

Moreover, this research only made a general distinction between monetary measures, traffic regulations, and infrastructure measures. As the comparative analysis of the 20 countries shows, various differences exist in the exact types of incentives that governments have implemented. Future research should investigate in detail which types of policy measures are the most influential for promoting EV penetration. In particular, having found that monetary measures and charging infrastructure investments are the most effective incentives, focusing on these measures in future research may be a good means to produce findings that are even more actionable.

Finally, it is important to take into account that this research only represents a snapshot of the current EV market, using data from 2016. Since governmental incentives and the market itself have changed significantly over the past several years and are likely to continue to develop in the future, it is important to track these changes over time and continuously reconsider the current findings in light of these developments. In this way, it will be possible to provide valuable recommendations for the promotion of EVs among consumers over time.

References

Bakker S, Maat K, van Wee B (2014) Stakeholders interests, expectations, and strategies regarding the development and implementation of electric vehicles: the case of the Netherlands. Transp Res Part A Policy Pract 66(1):52–64. https://doi.org/10.1016/j.tra.2014.04.018

Bakker S, Trip JJ (2013) Policy options to support the adoption of electric vehicles in the urban environment. Transp Res Part D Transp Environ 25:18–23. https://doi.org/10.1016/j.trd.2013.07.005

Bakker S, Trip JJ (2015) An analysis of the standardization process of electric vehicle recharging systems. In: Filho WL, Kotter R (eds) E-mobility in Europe: trends and good practice. Springer, Cham, pp 55–71

Baran R, Legey LFL (2013) The introduction of electric vehicles in Brazil: impacts on oil and electricity consumption. Technol Forecast Soc Chang 80(5):907–917. https://doi.org/10.1016/j.techfore.2012.10.024

Beltramello A (2012 Sept) Market development for green cars. OECD Green Growth Pap 3:158. https://doi.org/10.1787/5k95xtcmxltc-en

Benvenutti LMM, Ribeiro AB, Forcellini FA, Maldonado MU (2016) The effectiveness of tax incentive policies in the diffusion of electric and hybrid cars in Brazil. In: 41st Congresso Latinoamericano de Dinamica de Sistemas, São Paulo

Berman B (2017) Incentives for plug-in hybrids and electric cars. Plugincars. http://www.plugincars.com/federal-and-local-incentives-plug-hybrids-and-electric-cars.html. Accessed 27 Apr 2017

Bundesamt für Energie (2017) Kantonale Motorfahrzeugsteuern: Rabatte für energieeffiziente Fahrzeuge. http://www.bfe.admin.ch/energieetikette/00886/02038/index.html?lang=de&dossier_id=02083. Accessed 4 Apr 2017

ChargeHub (2017) Charging stations. https://chargehub.com/en/charging-stations-map.html. Accessed 27 Apr 2017

Cohen J (1988) Statistical power analysis for the behavioral sciences, 2nd edn. Lawrence Erlbaum Associates, Hillsdale

Domingues JM, Pecorelli-Peres LA (2013) Electric vehicles, energy efficiency, taxes, and public policy in Brazil. Law Bus Rev Am 19(1):55–80

EAFO (2017a) Incentives & legislation. http://www.eafo.eu/incentives-legislation. Accessed 6 Apr 2017

EAFO (2017b) Electric vehicle charging infrastructure. http://www.eafo.eu/electric-vehicle-charging-infrastructure. Accessed 26 Apr 2017

Environmental Protection Department (2017) Promotion of electric vehicles in Hong Kong. http://www.epd.gov.hk/epd/english/environmentinhk/air/prob_solutions/promotion_ev.html Acces sed 3 Apr 2017

European Commission (2018) Reducing CO_2 emissions from passenger cars. https://ec.europa.eu/clima/policies/transport/vehicles/cars_en. Accessed 1 Feb 2018

EV-Volumes (2017) EV-Volumes. http://www.ev-volumes.com/. Accessed 20 June 2017

Figenbaum E, Assum T, Kolbenstvedt M (2015) Electromobility in Norway—experiences and opportunities. Res Transp Econ 50:29–38. https://doi.org/10.1016/j.retrec.2015.06.004

Gazeta R (2016) Subsidizing purchases of electric cars would make perfect sense. Analytical Center for the Government of the Russian Federation. http://ac.gov.ru/en/commentary/09599.html. Accessed 4 Apr 2017

Government of NCT of Delhi (2016) Budget 2016–2017. http://delhi.gov.in/wps/wcm/connect/DoIT_Planning/planning/budget+of+delhi/budget+2016-17. Accessed 20 Apr 2017

Haddadian G, Khodayar M, Shahidehpour M (2015) Accelerating the global adoption of electric vehicles: barriers and drivers. Electr J 28(10). https://doi.org/10.1016/j.tej.2015.11.011

Hao H, Ou X, Du J, Wang H, Ouyang M (2014) China's electric vehicle subsidy scheme: rationale and impacts. Energy Policy 73:722–732. https://doi.org/10.1016/j.enpol.2014.05.022

Holtsmark B, Skonhoft A (2014) The Norwegian support and subsidy policy of electric cars. Should it be adopted by other countries? Environ Sci Policy 42:160–168. https://doi.org/10.1016/j.envsci.2014.06.006

IEA IA-HEV (2008) Hybrid and electric vehicles: the electric drive gains momentum. http://www.ieahev.org/news/annual-reports/

IEA IA-HEV (2011) Hybrid and electric vehicles: the electric drive plugs in. http://www.ieahev.org/news/annual-reports/

IEA IA-HEV (2012) Hybrid and electric vehicles: the electric drive captures the imagination. http://www.ieahev.org/news/annual-reports/

IEA IA-HEV (2013) Hybrid and electric vehicles: the electric drive gains traction. http://www.ieahev.org/news/annual-reports/

IEA IA-HEV (2014) Annual report 2013: hybrid and electric vehicles—the electric drive accelerates. http://www.ieahev.org/news/annual-reports/

IEA IA-HEV (2016) Annual report 2015: hybrid and electric vehicles—the electric drive commutes. http://www.ieahev.org/news/annual-reports/

JAMA (2010a) The motor industry of Japan 2010. http://jama-english.jp/publications/MIJ2010.pdf

JAMA (2010b) Japanese government incentives for the purchase of environmentally friendly vehicles. Japan Automobile Manufacturers Association. http://www.jama.org/japanese-government-incentives-for-the-purchase-of-environmentally-friendly-vehicles/

Kim S, Yang Z (2016) Promoting electric vehicles in Korea. International Council on Clean Transportation (ICCT). http://www.theicct.org/blogs/staff/promoting-electric-vehicles-in-korea

Li Y (2016) Infrastructure to facilitate usage of electric vehicles and its impact. Transp Res Procedia 14:2537–2543. https://doi.org/10.1016/j.trpro.2016.05.337

Lieven T (2015) Policy measures to promote electric mobility—a global perspective. Transp Res Part A Policy Pract 82:78–93. https://doi.org/10.1016/j.tra.2015.09.008

Marchán E, Viscidi L (2015) The outlook for electric vehicles in Latin America. http://www.thedialogue.org/wp-content/uploads/2015/10/Green-Transportation-The-Outlook-for-Electric-Vehicles-in-Latin-America.pdf

Maurer J (2014) Taiwan setzt bei Elektromobilität Prioritäten. Germany Trade & Invest (GTAI). http://www.gtai.de/GTAI/Navigation/DE/Trade/Maerkte/suche,t=taiwan-setzt-bei-elektromobilitaet-prioritaeten,did=1081172.html. Accessed 5 Apr 2017

Mock P, Yang Z (2014) Driving electrification—a global comparison of fiscal incentive policy for electric vehicles. International Council on Clean Transportation (ICCT). http://www.theicct.org/sites/default/files/publications/ICCT_EV-fiscal-incentives_20140506.pdf

NationMaster (2017) Motorway length by country. http://www.nationmaster.com/country-info/stats/Transport/Road/Motorway-length. Accessed 5 May 2017

Netherlands Enterprise Agency (2017) Electric transport in the Netherlands. https://www.rvo.nl/sites/default/files/2017/04/Highlights-2016-Electric-transport-in-the-Netherlands-RVO.nl_.pdf

OECD/IEA (2016) Global EV outlook 2016: beyond one million electric cars. International Energy Agency (IEA). https://www.iea.org/publications/freepublications/publication/global-ev-outlook-2016.html. Accessed 18 Apr 2017

Office of Energy Efficiency & Renewable Energy (2018) Electric vehicles: tax credits and other incentives. https://energy.gov/%0Aeere/electricvehicles/electric-vehicles-tax-credits-and-other-incentives%0A. Accessed 5 Jan 2018

OICA (2017a) Sales of new vehicles 2005–2016. Organisation Internationale des Constructeurs d'Automobiles (OICA). http://www.oica.net/category/sales-statistics/. Accessed 20 June 2017

OICA (2017b) 2016 Production statistics. Organisation Internationale des Constructeurs d'Automobiles (OICA). http://www.oica.net/category/production-statistics/2016-statistics/. Accessed 15 Nov 2017

PlugShare (2017) EV charging station map. https://www.plugshare.com/#. Accessed 26 Apr 2017

Prefeitura de São Paulo (2015) Veículos elétricos e híbridos poderão ter desconto de 50% do IPVA. http://capital.sp.gov.br/noticia/veiculos-eletricos-e-hibridos-poderao-ter-desconto. Accessed 20 Apr 2017

PwC (2016) 2016 Global automotive tax guide. https://www.pwc-wissen.de/pwc/de/shop/publikationen/Global+Automotive+Tax+Guide/?card=20976

Rokadiya S, Bandivadekar A (2016) Hybrid and electric vehicles in India: current scenario and market incentives. http://www.theicct.org/sites/default/files/publications/India-hybrid-and-EV-incentives_working-paper_ICCT_27122016.pdf

Selya AS, Rose JS, Dierker LC, Hedeker D, Mermelstein RJ (2012) A practical guide to calculating Cohen's f2, a measure of local effect size, from PROC MIXED. Front Psychol 3: 1–6. https://doi.org/10.3389/fpsyg.2012.00111

Sierzchula W, Bakker S, Maat K, Van Wee B (2014) The influence of financial incentives and other socio-economic factors on electric vehicle adoption. Energy Policy 68:183–194. https://doi.org/10.1016/j.enpol.2014.01.043

Teixeira ACR, da Silva DL, de Neto LVBM, Diniz ASAC, Sodré JR (2015) A review on electric vehicles and their interaction with smart grids: the case of Brazil. Clean Technol Environ Policy 17(4):841–857. https://doi.org/10.1007/s10098-014-0865-x

The Norwegian Tax Administration (2017) Pay annual motor vehicle tax. http://www.skatteetaten.no/en/person/cars-and-other-vehicles/annual-motor-vehicle-tax/pay/. Accessed 16 Nov 2017

The World Bank (2017) GDP per capita (current US$). The World Bank. https://data.worldbank.org/indicator/NY.GDP.PCAP.CD. Accessed 15 November 2017

van't Hull C, Linnenkamp M (2015) Rolling out e-mobility in the MRA-electric region. In: Filho WL, Kotter R (eds) E-mobility in Europe: trends and good practice. Springer, Cham, pp 127–140

van der Steen M, van Schelven RM, Kotter R, van Twist MJW, van Deventer P (2015) EV policy compared: an international comparison of governments' policy strategy towards e-mobility. In: Filho WL, Kotter R (eds) E-mobility in Europe: trends and good practice. Springer, Cham, pp 27–53

Worldometers (2017) World population by country. http://www.worldometers.info/world-population/. Accessed 25 Apr 2017

Zhang Y, Yu Y, Zou B (2011) Analyzing public awareness and acceptance of alternative fuel vehicles in China: the case of EV. Energy Policy 39(11):7015–7024. https://doi.org/10.1016/j.enpol.2011.07.055

Zhu G, Hein CT, Ding Q (2017) Case study—China's regulatory impact on electric mobility development and the effects on power generation and the distribution grid. In: Liebl J (ed) Grid integration of electric mobility: 1st international ATZ conference. Springer, Wiesbaden, pp 13–29

Nele Rietmann is a Ph.D. student in marketing at the University of St. Gallen, Switzerland. She completed her studies in management and marketing at the University of St. Andrews, Scotland, and at the University of St. Gallen. Since 2017, she has worked at the Institute for Customer Insight and conducted research on consumer behavior in the area of electric mobility. Her research with Prof. Dr. Theo Lieven includes analyzing the effectiveness of policy measures on the market shares of electric vehicles, which she presented at the 2017 International Conference on Urban Transport and the Environment, as well as the 2018 Annual Meeting of the Transportation Research Board. Further research investigates effective advertising strategies for electric vehicles.

Theo Lieven founded a computer company in 1975, which grew to be the largest European retailer by the mid-1990s. After the sale of the company, he studied management and economics and received his doctoral degree in marketing at the University of St. Gallen, Switzerland. He finished his habilitation in 2015 and in 2017 was appointed adjunct professor of marketing at the Institute for Customer Insight, University of St. Gallen. Recently, Palgrave Macmillan published his book about brand gender and the positive effect of brand personality on brand equity. He is mainly interested in public policies, particularly in the field of transportation and electric mobility.

Chapter 8
Nurturing a Regime Shift Toward Electro-mobility in Norway

Marianne Ryghaug and Tomas Moe Skjølsvold

Abstract Sales of electric vehicles (EVs) have exploded over the past several years in Norway, to the point that new EVs now outnumber new gas-driven cars in current sales. The popular narrative about how this transition came about suggests that it was the result of a targeted set of policies aiming to stimulate demand for EVs. In this chapter, we tell a different story. In looking at the history behind these ambitious policies, we aim to show that the policies were originally implemented to stimulate the development of a Norwegian EV industry. During the 1990s, much work was done among various industrial actors, NGOs, and policy-makers to establish a new Norwegian niche industry venture, which was partially inspired by local policies implemented in California. The venture did not come to fruition, but the policies eventually did, together with changes in mobility culture, creating one of the world's strongest EV markets. The story illustrates the importance of understanding not only how policies work, but also how they are produced and how their effects travel across geographical borders.

Keywords Electro-mobility · Transition · Policy · Scale · Culture

8.1 Introduction

In 2010, approximately 3000 battery electric vehicles (EVs) could be seen on Norwegian roads, and their sales were hardly visible in annual statistics. As we write this chapter, eight years later, almost every other new car sold in Norway is a battery EV, and the total market share is approximately 30%. EVs have become mainstream and are normalized elements in Norwegian mobility culture. Their

M. Ryghaug (✉) · T. M. Skjølsvold
Department of Interdisciplinary Studies of Culture, Norwegian University of Science and Technology (NTNU), Trondheim, Norway
e-mail: marianne.ryghaug@ntnu.no

T. M. Skjølsvold
e-mail: tomas.skjolsvold@ntnu.no

M. Finger and M. Audouin (eds.), *The Governance of Smart Transportation Systems*, The Urban Book Series, https://doi.org/10.1007/978-3-319-96526-0_8

established presence suggests that it is now possible to speak about a transition and raises the question as to how we might understand these developments.

The story of a Norwegian EV transition has received substantial international interest. The standard media narrative sees the Norwegian EV boom as a result of targeted policies aiming to stimulate EV demand. This chapter offers a counter-narrative by exploring the history of the Norwegian EV transition in depth, introducing important nuances to the story and posing some challenges to the transition framework, as explicated through the first generation of multi-level perspective (MLP). On the one hand, we illustrate that many of the incentives that seem to underpin the current boom in EV demand were, in fact, introduced more than a decade ago and, in some instances, as much as 20 years before sales figures peaked. On the other hand, we show that the primary objective of these incentives was not to stimulate mass-market demand, but to nurture what many hoped would be the next Norwegian industrial venture: production and export of Norwegian EVs. The quest for such an ambitious industrial undertaking was partially fueled by local policies in California and subsequent industrial strategies adopted by international incumbents to meet new local regulations.

In the mid-1990s, a Norwegian transportation researcher concluded that Norwegian EV policies were a failure (Buland 1994), echoed in the international literature (Hoogma et al. 2002: 184), stressing that the small Norwegian market emerged from *distinct and specially created circumstances [...] that cannot easily be copied to other countries*. In this chapter, we explore the transition, more than 20 years later, when many of the same policies are making international headlines, and Norwegian authorities are being praised for the success of the electric car market.

Currently, EVs in Norway are part of a narrative of climate mitigation. While the environmental merits of EVs are sometimes contested in the media and popular debate, life-cycle analysis indicates that EVs have substantial climate benefits in European contexts, even in settings where electricity is produced by coal or gas (Hawkins et al. 2013; Ellingsen et al. 2016). In Norway, this positive effect is even greater because Norwegian electricity production is predominantly renewable (98%) and based on hydropower (e.g., Skjølsvold et al. 2013). Therefore, the low-hanging fruit of Norwegian climate mitigation is not to reduce fossil fuels in electricity generation (as is the case for many other countries), but to electrify the transportation sector (Aamaas and Peters 2017).

Against this backdrop, the story of Norwegian policies intuitively boosting demand for EVs makes sense and is strengthened by the fact that Norway is a particularly mass motorized society (Østby 2004). Living standards and wages are high, with a "comfort-oriented" energy culture, in which electricity is both abundant and cheap (Aune 2007). Retrospectively, the EV appears to be a natural fit for Norway's national context. However, factors such as the large Norwegian export of oil and gas, and its importance for Norway's GDP, might lead us to conclude that promoting transportation electrification is misaligned with incumbent oil and gas interests.

The Norwegian mobility transition does not entail reduced car sales. The year 2016 ranked third for car sales in Norway, with 154,603 new private cars registered, which was a 2.6% increase from 2015. Approximately 30% of these were EVs or chargeable hybrid EVs. Another 10% were nonchargeable hybrid EVs. Of all EVs sold in Western Europe in 2014, 35% were sold in Norway. By September 2017, the sales of new electric passenger cars continued to grow, reaching a record-high 28.6% of the market share. If the current trend continues, the share of electric cars will continue to grow.

Although the EV share is not higher than 3.7% of the total number of cars in Norway, the country has taken a leading role in introducing electric cars, acting as a kind of laboratory for experiments in developing a market for EVs. Consequently, interest in the Norwegian experience has been high among international analysts and practitioners working with electrification. As noted, this interest has clustered around policies. There has been a strong political drive to reduce greenhouse gas in the Norwegian transportation sector. A comprehensive package of local economic incentives, as well as the establishment of a state-owned enterprise called Transnova (now merged with Enova), which provides financial support for charging facilities, was important for rapidly expanding Norwegian EV sales (Figenbaum and Kolbenstvedt 2013; Ryghaug and Toftaker 2014). Incentives to promote EVs in Norway include exemptions from sales tax, vehicle registration, and value-added tax (VAT). Furthermore, electric cars are except from road tolls and tunnel-use charges, granted reduced fares on ferries, can use bus lanes, benefit from public parking (sometimes with free charging), and have access to a dispersed network of charging stations. This appears to be a solid package to stimulate EV demand.

One aspect of the package that has received substantial interest is its effects on the price of EVs. An electric car in Norway is typically priced in the same range as a gas-driven car in the same class (in other words, the electric version of a VW Golf costs almost the same as its gas-driven counterpart and benefits from tax reductions, ranging from €7000 to €8000). Furthermore, operational costs of electric cars are relatively low, due to effective engines fueled by cheap electricity produced by hydropower. The total savings of driving EVs depend on a variety of factors (such as driving style, and use of toll roads and ferries), but the cost of fuel (electricity) is about one-fourth to one-fifth of the cost for petrol. For instance, driving a Nissan Leaf, with an annual mileage of 15,000 km, costs about €2800 less annually than a comparable gas-driven car.

In a white paper from 2012, the Norwegian government stated that the comprehensive package of electric car incentives would be prolonged until either 2017 or the number of EVs rose to 50,000. As this objective was reached in 2015, the incentives have been widely debated since then. In December 2016, Norway had 100,000 EVs, several years earlier than expected. The incentives will be revised and adjusted in parallel with market development in the years to come, but the government promised to keep the tax incentives until at least 2018. However, the ambitions are still high as seen by the Norwegian Parliament's goal that all new cars sold by 2025 should be either zero emission (electric or hydrogen) or low emission (plug-in hybrids).

As we have seen, there are strong incentives in Norway for purchasing and driving EVs, but the common story of these incentives making Norway the global forerunner in electro-mobility (Bjerkan et al. 2016) is too simplistic. In fact, the Norwegian story illustrates that economic incentives alone cannot explain or ensure large sales. Many economic incentives were introduced during the 1990s or early 2000s, without any significant effects on the market: EVs remained a niche market (Figenbaum et al. 2015). Viewed this way, we might widen our perspective and look at the role of other changes. The technological development of electric cars, particularly their battery technology, is one very tangible aspect. Furthermore, there are more subtle changes in how Norwegians talk, think, and act with respect to mobility. In the words of Sheller (2014), it is possible to observe not only a technological transition, but a transition of practices, networks, and discourses; in other words, an unfolding transition of mobility culture (see also, Hopkins and Stephenson 2014).

The analysis is based on a compilation of findings from a number of different research projects studying different aspects of electrification of the Norwegian transportation sector in which the authors have been involved. These projects yielded numerous interviews with both users and key figures involved in the introduction of EVs in Norway, as well as document analysis. Empirically, the chapter is based on data from these previous studies, official transportation policy documents, as well as available secondary sources, such as journal articles and books.

The chapter is structured as follows: Sect. 2 introduces the multi-level perspective (MLP). Section 3 applies MLP to the Norwegian electro-mobility system and analyses of the dynamics between policies, actors, and market development across time and space. This empirical application has the character of an interpretive assessment, with trade-offs between breadth and depth. While the assessment is broad, in order to address various dimensions of the electro-mobility system and change initiatives, many nuances and complexities must be relegated to the background in favor of larger patterns, strategic decisions, and important events. The discussion also highlights some challenges to the MLP framework identified through the analysis. Section 4 draws conclusions about low-carbon transitions and makes some evaluative remarks on what other countries can learn from Norwegian electro-mobility experiences.

8.2 A Socio-technical Transition Perspective

Systemic transitions entail coevolution and multi-dimensional interactions between industry, technology, markets, policy, culture, and civil society (Geels 2012). To understand the development and increasing proliferation of EVs in Norway, we used a socio-technical understanding anchored in the MLP (e.g., Geels et al. 2014; Geels 2010). This entails a symmetrical understanding of the importance of social and technical elements of transitions and recognition that the elements in socio-technical systems are maintained, reproduced, and changed by various actor groups (Geels 2012). Transitions are coevolutionary processes that take decades to unfold and involve many actors and social groups (e.g., firms and industries, policy-makers and politicians, consumers, civil society, engineers, and researchers). MLP further distinguishes between three levels: niches, regimes, and landscapes (Geels 2002). The top-level landscape is exogenous to the system. It is the *technical, physical, and material backdrop that sustains society* (Geels and Schot 2007: 403). Change is very slow, with the exception of external shocks. Regimes are constructed of stable, institutionalized, large networks, while niches are smaller, with less stabilized rules of conduct.

The model of agency in the MLP builds on institutional theory (Scott 1995) and sociological structuration theory (Giddens 1984), which implies that actors in regimes and niches make choices under the influence of regulatory, cognitive, and normative rules (see Geels 2010). These rules guide actors, who also produce and reproduce the rules through their enactment. The landscape agency does not necessarily determine what happens in regimes, but *provides deep-structural "gradients of force" that make some actions easier than others* (Geels and Schot 2007: 403). Transitions are changes in the regime, often enabled by nurturing niche technologies and solutions to eventually grow into, and destabilize the regime. A recurring issue in such transition processes is that there is a lack of coherence between the societal institutions, or the rules of the game, and the technologies being implemented. As an example, it is quite common for institutions to be shaped for centralized systems, while emerging systems are more distributed (Crettenand and Finger 2013).

Although we recognize that the MLP offers many clues about how to understand and analyze long-term, encompassing transitions, we are also sensitive to criticism that MLP focuses too much on the semi-functionalistic aspects of systems, and not enough on the actors involved in transitions and their practices (Åm 2015; Farla et al. 2012; Smith and Raven 2012). We also support, and build on, recent attempts to better understand the formation of policy processes leading to transitions (Kern and Rogge 2017), as to how different network structures facilitate different levels of access to the policy-making process (Normann 2015). It will be central for us to discuss not only how policies work as a factor influencing EVs diffusion, but more fundamentally, how and why policies have been shaped in the way they have. Furthermore, we are interested in the relationship between policies implemented in diverse geographic locations and processes unfolding across countries and continents in unexpected ways.

Another important aspect of a socio-technical perspective on transitions in mobility is a renewed interest in the cultures of mobility, the elements that constitute such cultures, and the roles of these cultures in mobility transitions (e.g., Hopkins and Stephenson 2014; Sheller 2012). Sheller's (2012) contribution is particularly interesting for our discussion. Sheller highlights how the niches, regimes, and landscapes of mobility are all produced by a set of three distinct elements: practices, networks, and discourses. Understanding the journey of a new solution, such as the EV transitioning from niche to regime, is not only a matter of understanding the proliferation of the technological artifact, or related infrastructural elements such as filling stations and repair shops. The shift from niche to regime phenomenon posits that practices change from embodying alternative subcultural mobilities to mainstream legitimized practices. Networks shift from being those of social movements rooted in green lifestyles to those of durable interest groups and governing structures, while discourses shift from counter-discourses that challenge dominant order to standard discourses that legitimize existing actors and practices. Sheller's research adds further analytical depth to what Hopkins and Stephenson (2014) call mobility cultures, which are created out of materiality, cognitive norms, and social practices. Our discussion emphasizes the relationships between social and material aspects of electro-mobility, and the sometimes unexpected links that emerge between policy, practice, innovation, and diffusion that we see as decisive in the shift toward electro-mobility in Norway. Some of these links were already apparent at the beginning of the century, when Gjøen and Hård (2002) noted that by driving differently and viewing automobility differently, EV owners developed user scripts that challenged established political and engineering scripts, while contributing to a cultural politics of automobility.

8.3 Analysis: Nurturing a Norwegian Mobility Regime Shift?

Discussions about transitions often revolve around how to nurture niche industries, socio-technical configurations, and technologies for them to flourish and gain a foothold at the regime level (Geels 2002). This somewhat broader consideration at the regime level aptly suggests that transitions are about more than simply transplanting new technologies into social settings, but that they are also about producing new industries, business, practices, and culture. In this section, we will analyze why Norway embarked on this particular transition pathway from traditional fossil-fuel cars to EVs when, at first glance, nurturing an electric EV market appears to be a poor match with domestic industry interests heavily entrenched in an oil economy, with no EV industry of which to speak. To understand Norwegian policy developments and governance structures in this area, we must first look back several decades and focus on a lesser-known aspect of its EV story—Norwegian car manufacturing and efforts to develop a domestic EV industry.

8.3.1 Early Attempts to Nurture an Alternative EV Industry in Norway

Norway launched several initiatives to develop electric cars and engaged in multiple initiatives to launch and develop a motorized vehicle industry. Two Norwegian pioneer cars were developed in 1895 and 1896: the Irgens and the Vestby (but the companies were short-lived). Subsequent attempts at launching an automobile industry also failed. The car manufacturing company, Troll, sold its first car in 1956, but went bankrupt in 1958, having delivered only six cars. EVs were also produced in Norway between 1918 and 1924. The production company, Staværn Bilfabrikk, was created to mitigate the problems of obtaining motorized vehicles after WWI. The company delivered 10 functional electric trucks (Asphjell et al. 2013).

During the 1970s, interest in EVs rose sharply in response to the oil crisis of 1973. A company called ELBIL (which literally translates into electric car) delivered three electric vans to state service providers (Asphjell et al. 2013: 52). However, the most important development in this period was that the owners of Bakelittfabrikken AS, a plastic industrial firm, aimed to produce a small, urban, plastic-chassis EV. This strategy was based on the notion that Norway was poor in oil but wealthy in electricity, which should be reflected in the country's dominant mode of mobility (Asphjell et al. 2013). A prototype was built, but no subsequent steps were taken.

The developments of the 1930s and 1970s illustrate how landscape shocks, such as the oil crisis, might open windows of opportunity for new niche transportation technologies (Geels and Schot 2007). However, the dominant automobility regimes remained intact after the oil crisis ended, and interest in EVs decreased. After some initial work in the late 1980s, the owners of Bakkelittfabrikken AS started a new company, called Personal Independent Vehicle Company (PIVCO) in 1990. The idea was nurtured through funding from the Natural Sciences Research Council of Norway, resulting in a feasibility study published around the time of the company's founding (Røste 2001). The study's practical outcome was the ambition to build a short-range, two-seat EV, called a personal independent vehicle (PIV) (Buland 1994). This rekindled interest in electric mobility was also inspired by events on the other side of the world. Enactment of the Zero Emission Vehicle (ZEV) legislation in California offered future commercial opportunities (Hoogma et al. 2002; Buland 1994).

Bakelittfabrikken was an opportunity to create new business and industrial opportunities in Norway to compete against the comparatively larger Swedish automotive industry. The firm secured loans and government subsidies, attracted interest and support from a significant number of private and public actors, and obtained R&D funding from various sources (Hoogma et al. 2002). The first prototype (PIV1) was successfully tested in 1993, resulting in a new project for which PIVCO delivered a fleet of 13 EVs (PIV2) to be tested in extremely cold conditions. These EVs garnered a great amount of public visibility in its trials during the 1994 Winter Olympic Games in Lillehammer (Asphjell et al. 2013). The PIV2 was re-branded as the CityBee for these trials. The Lillehammer demonstration was surprisingly successful. On the one hand, it was a niche experiment (Raven et al. 2012). On the other hand, it was spectacularly visible, functioning as both a marketing activity and a public-engagement activity to illustrate an alternative to the dominant mobility regime. PIVCO's work at that time was subsidized by funding from a national industrial fund, as well as supported by Oslo Energi, a large

Oslo-based electricity producer (Røste 2001). Some level of national nurturing and protection of this small, niche product was required.

The CityBee experiment demonstrated the vehicles' potential and attracted the interest of several incumbent actors in Norway and abroad. Local electric companies Oslo Energi, Østfold Energi, and Stavanger Energi were all early customers. These companies were interested in using the vehicles for marketing purposes and showcasing various uses of electricity (Buland 1994). It is important to note that these actors were not from the traditional automobile industry, so the experiment did not belong to the international automobile-production regime. They were mainly part of the electricity-production sector, which was an alternative regime.

Meanwhile, PIVCO attracted international interest. San Francisco was developing its profile as a pioneer of clean, urban transportation and was looking to supplement its Bay Area Rapid Transportation System (BART) light rail system. The city ordered approximately 50 vehicles from PIVCO for its collective station car program (Asphjell et al. 2013: 127), in order to challenge a strong cultural preference for personal car ownership (Geels 2012). It is likely that this initiative was made possible by combining actors from two different kinds of regimes: one was traditionally involved in providing hydropower and electricity in Norway, and the other was involved in rail-based public transportation in San Francisco. Together, they enabled production of a distinctly new kind of car with a different ownership structure, script, and intended use than that of traditional cars. PIVCO's development surged in 1995, with several large publicity stunts in support of EVs in Norway. When delivering the first vehicles to San Francisco, PIVCO management was escorted by the Norwegian king and queen, securing them massive media attention. The first Scandinavian electric car rally, from Gothenburg to Oslo, was hosted the same year, including famous Scandinavian rally drivers. Norwegian actors in the EV sector was that Norway was about to embark on a new, widespread EV industrial venture.

After some difficult years requiring intensive work in San Francisco, PIVCO's entry into the USA aroused substantial interest from the traditional automobile regime. This interest was amplified by the Zero Emission Vehicle legislation in California, which established a credit system in which car dealers must earn credits from selling non-emission vehicles to legally continue selling gas cars (Hoogma et al. 2002). Actors like Chrysler and General Motors took legal action against the state, but Ford was determined to comply with the new rules, opening up a new window of opportunity for Norwegian EV manufacturers. Ford acquired PIVCO in order to meet the new California legislation requirements. By the late 1990s, PIVCO was re-branded as Th!nk. The company that had been nurtured and assisted by work in alternative regimes was now appropriated and made part of a traditional automobile-production regime. This shift entailed large changes for Th!ink, which had to adjust to Ford's production standards—not only upscaling but also changing how vehicles were produced, with much higher performance expectations. In the eyes of many Norwegians, the EV adventure had now come to fruition. The standard narrative highlights how Ford's *massive automotive competence* was what PIVCO had been missing (Røste 2013: 7). However, in retrospect, an equally plausible interpretation is that Ford's acquisition of Th!nk was the beginning of the end for the Norwegian EV industry. We will return to this point, but let us first look at some other parallel developments.

Ford's acquisition of Th!nk in 1999, and the subsequent launch of the first model intended for mass marketing, has been described by transportation scholars as the early market phase of Norwegian EV development (Figenbaum and Kolbenstvedt 2013). At the same time, there were other actors creating new companies to become part of the Norwegian EV venture. Kollega Bil was established and started producing and leasing the EV brand Kewet in Norway after buying the assets from a bankrupt estate in Denmark (Figenbaum and Kolbenstvedt 2013). Other external factors were also favorable during this period. The big industrial conglomerate Norsk Hydro had to scale down its activities in the region, resulting in more extensive business development support, which also benefited Miljøbil Grenland's new EV-leasing business operating in the area. Consequently, a Norwegian EV industry cluster was in the making, as was the political understanding that it was important to support the development of a domestic EV market (Figenbaum and Kolbenstvedt 2013). As Gjøen and Hård (2002) noted, politics were not only conducted through formal processes, but also through distributed processes of micropolitics, in which strategies of actors, such as municipalities and individual drivers, were important.

The Norwegian EV adventure was nurtured in several ways. First, there were small-scale national funding mechanisms meant to protect PIVCO and accelerate the industrial evolution of the company. Second, there were local policy initiatives abroad, notably in San Francisco and the state of California. Third, actors promoting electro-mobility began coordinating and organizing their actions in a targeted way in the early 1990s. The EV interest organization NORSTART was established in the early 1990s, aiming to pressure the government and unify what was still an uncoordinated business area (Buland 1994). The organization was quite successful, and several incentives to stimulate the demand for EVs were introduced as the story of PIVCO and the Norwegian EV industrial adventure unfolded. EVs were exempt from sales and import taxes in 1990. Some places implemented free parking in 1993, and most municipalities had free parking starting in 1999. EVs benefited from low annual road taxes starting in 1996 and were exempt from toll roads in 1997. NORSTART was not the only actor behind these policy developments. The environmental NGO Bellona, which worked to raise awareness of EV benefits, must also be credited for its long-standing effort to secure favorable conditions for EVs in Norway, particularly in Oslo. With the emergence of a new Norwegian EV industry cluster, stimulating the development of a domestic market was important. The result was a set of new incentives: exemption from VAT, starting in 2001 (25%); experiments with allowing EVs to drive in bus lanes in the greater Oslo region, starting in 2003 (permanent and nationwide, starting in 2005, with minibuses banned, starting in 2009); and reduced rates on coastal ferries (starting in 2009), and exemption from VAT on leasing (starting in 2015).

Despite *wide-ranging political visions, far-reaching networks, and elaborate engineering scripts*, the number of EVs was still limited in 2002 (Gjøen and Hård

2002). However, after more than 10 years of attempting to establish domestic manufacturing, the efforts to promote EV technology began to pay off. Buland (1994) asked if a lack of tradition for car manufacturing actually could be beneficial for producing a new electric car in Norway. The question is still relevant. The Norwegian EV adventure was mainly driven by actors with no prior interest or competence in car production. They were not restricted to a set of predefined car models and existing sociocultural understandings about what a car was or could be. Nor were they restricted by existing manufacturing techniques and the so-called sunk investments related to production modes and facilities, or networks of existing interests. When PIVCO was approached by San Francisco, there was no threat to regime ideals about producing and selling cars to individuals. It might not be so strange that the EV challenge to traditional automotive regimes emerged from Norway, as a country without a strong car manufacturing tradition.

8.3.2 The Harsh Reality of the International Automobile Regime and Its Fatal Consequences for the Norwegian EV Adventure

By 2003, it was clear that Chrysler and General Motors had won the lawsuit against the state of California, so ZEV regulations became weaker. This, combined with poor corporate economy, resulted in Ford pulling out of Th!nk. Compared to other cars in the same price range, the Th!nk car was small and relatively slow, making it difficult to introduce to the American market. One explanation was that Th!nk was a poor match with American mobility culture, which remained stable and anchored in hegemonic ideals of personal ownership of large gas-driven cars. Th!nk was eventually acquired by other investors, who owned the company for two years without achieving much. The company was again bankrupt in 2004. This time Th!nk was bought by Norwegian investors who wanted to revitalize the company by launching a new model developed during the period of Ford ownership. The domestic Norwegian EV market was relatively stagnant in this period. What little demand that existed was not covered by Norwegian industry, but by secondhand imports of French EVs manufactured between 1998 and 2002. The main EV market was located in the greater Oslo/Akershus region where commuters could save time, driving in the bus lanes and areas with high toll-road charges (Figenbaum and Kolbenstvedt 2013). According to Figenbaum and Kolbenstvedt (2013), the market introduction phase started around 2009, when a new generation of Th!nk was launched by new owners, and the alternative Norwegian brand, Pure Mobility (which produced the Buddy and Kewet models), surfaced. From 2010 to 2011, industry leaders Mitsubishi, Peugeot, Citroën, and Nissan began to launch EV models, and Norwegian car dealers began importing them. Norwegian EV manufacturers soon went bankrupt. The Norwegian EV market really boomed after the introduction of the Mitsubishi i-MiEV in 2010 and Nissan LEAF in 2011 (Lorentzen et al. 2017).

International discussions about the Norwegian EV boom focused on the period after 2009, which is not so strange, as it represents the first period in which a jump in sales statistics can be observed. In our analysis, this period was less significant because the introduction of new incentives and government support was limited. However, one important development was NORSTART becoming the EV Association, which entailed much stronger coordination efforts and much more active efforts to enroll the Norwegian public as participants in the EV transition. NORSTART disseminated information (such as on charging infrastructure), recruited EV drivers with free test drives, and facilitated knowledge transfer through online platforms. Norway's first governmental support scheme for public charging infrastructure took place in 2009 to 2010 (Lorentzen et al. 2017), resulting in approximately new 1800 chargers (Schuko-point, household sockets). The government organization Transnova (later merged with Enova) was established to support testing and implementing climate-friendly technologies in transportation. Transnova ensured the coordination of (fast) charging infrastructure and supported development of charging facilities, resulting in a large network of charging stations across the country. As of June 2017, there were approximately 4400 publicly available Schuko-point and 2700 Type 2 point charging stations (Lorentzen et al. 2017). In 2015, Enova introduced a support scheme to cover Norwegian main roads with fast-charging stations every 50 km and support building fast chargers in municipalities with less than two fast chargers available, on a first-come first-served basis. In 2015 and 2016, Enova awarded 50.5 million kroner (kr) to support the construction of 230 fast chargers on a number of routes in Norway (Lorentzen et al. 2017).

Developing an accessible, dispersed charging network probably had an important symbolic effect, as it made the EV support strategy highly visible. Most studies show that EV drivers most often tend to charge their vehicles at home (97% on a daily or weekly basis, for those living in detached housing; 64% for those in apartment buildings). Some people charge at work (approximately 37%), while a few (approximately 15%) use public charging stations daily or weekly (Lorentzen et al. 2017; Norwegian EV owner survey 2017). However, a network of chargers throughout the country may be a culturally important safety net to mitigate everyday anxiety about vehicle range.

8.3.3 User Preferences and the Growing EV Market

The Norwegian EV transition should not be reduced to a tale of implementing effective policies. Rather, there have been important changes in how Norwegians talk, think, and act with respect to mobility during the last decades. It illustrates that the Norwegian EV transition is in an unfolding transition of mobility culture, including changed practices, networks, and discourses (e.g., Sheller 2014; Hopkins and Stephenson 2014). Perhaps the most important developments after 2009 have been the increase in social learning among drivers of EVs and the gradual development of a mobility culture, in which EVs work and are embedded in Norwegian mobility culture. Throughout the period discussed, there have been a few, but important, studies of user preferences related to EV driving in Norway. These studies point to the possibilities of reframing what constitutes a car, a task that historically has been difficult to achieve (Hård and Jamison 1997). Gjøen and Hård (2002) illustrated that EV driving contributes to undermining what were then very dominant ideas about automobile design, anchored in fossil-fuel-powered vehicles. EVs remained an incomplete innovation within an alternative automobility niche for years because of aspects relating to size, driving range, and comfort. However, studies of actual Norwegian driver experiences tend to produce different narratives (e.g., Gjøen and Hård 2002; Ryghaug and Toftaker 2014; Ingeborgrud and Ryghaug 2017), highlighting other qualities and stressing that EVs actually perform much better than expected compared to their fossil-fuel counterparts by being quieter; easier to operate (due to fast acceleration), park, and charge (due to charging at home or at work); receiving more positive reactions from others; and being seen as environmentally benign. The first EVs were mainly small, two-seat passenger cars with limited driving range branded as an environmental transportation device, rather than an ordinary car (Ryghaug and Toftaker 2014; Gjøen and Hård 2002). These compact EVs with limited range fit well with the city-car users' script. Most drivers were content with their cars' performance and had adopted their usage accordingly, viewing most features as assets. From here, it is difficult to pinpoint the exact emergence of new markets and user segments as they developed. However, user studies conducted over the last several years (Ryghaug and Toftaker 2014; Ingeborgrud and Ryghaug 2017) show how new user groups were attracted to the technology as it developed, and the EVs on the market began to resemble more traditional cars. EVs were also introduced in different public sectors, such as the postal service and home care. A qualitative leap was made with the development of the five-seat car. With Tesla and other luxury cars being developed, a new EV market offered vehicles for those who wanted higher-end EVs or families needing a bigger EV with a longer driving range (Ingeborgrud and Ryghaug 2017).

In stark contrast to the commonly perceived drawbacks of EVs, studies regarding Norwegian EV-user preferences stress the benefits of driving EVs and their embodied qualities: strong emphasis on good driving capabilities, comfort, and the experience of driving with a better conscience. EV driving in Norway seems to be culturally performative of environmental- and climate-related concerns, aligning with landscape changes related to cutting greenhouse emissions and pro-environmental actions. As an example, studies indicated that those who drive EVs are more likely to be interested in acquiring other environmentally oriented technologies such as solar panels (Throndsen et al. 2017; Ingeborgrud and Ryghaug 2017).

Why has Norway succeeded in increasing—more so than in other countries—its share of EVs compared to traditional cars? A broad set of incentives has been important, but detailed studies of actual EV users found that their understanding of the economic incentives varied (Ingeborgrud and Ryghaug 2017). For some, the incentives were important to promote initial adoption in the transition from gas to electric cars. For others, the driving pleasure related to EVs as green, nonpolluting cars was more important. The combination of economic and non-economic benefits represents a highly visible, concerted policy in support of EVs and has a dual effect. First, the comprehensive benefits provide instrumental motives to buy an EV. Second, the policy package clearly identifies EVs as a preferred alternative of policy-makers for a more sustainable mobility technology. In sum, we observe that there is an ongoing shift in the mobility culture, in which practices change from embodying alternative subcultural mobilities to mainstream legitimized practices (Sheller 2014) and networks change from social movements focused on alternative green lifestyles to more durable interest groups and governing structures (e.g., Ryghaug and Toftaker 2016). Discourses shift from being counter-discourses that challenge dominant stories to standard discourses that legitimize existing actors and practices. Table 8.1 summarizes some key findings from our discussion.

Table 8.1 Key events, dynamics, policies, and market developments in Norwegian EV transition

	Event	Key dynamics	Norwegian EV policies	Market
1970s	Proto PIVCO produced	Landscape shock: oil crisis	None	None
1990	PIVCO started	Inspiration: California ZEV legislation nurturing through research funds	None	None
1992	NORSTART, interest organization launched	Industry and interest coordination	Exemption from registration tax (1990)	Marginal niche market
1993	PIV1 tested	Backing from electricity-production regime	Free parking experiments (1993)	Marginal niche market
1994	Fleet of PIV2 (CityBee) demonstrated at Winter Olympics	Backing from electricity-production regime, niche experiment with large international audience	No new policies	Marginal niche market
1995	Fifty CityBees sold to San Francisco	Public transportation regime in San Francisco wants new solutions for transportation around light rail stations. PIVCO now anchored in Norwegian hydropower regime and San Francisco public transportation regime	No new policies	Marginal niche market in Norway, public transportation in USA

(continued)

Table 8.1 (continued)

	Event	Key dynamics	Norwegian EV policies	Market
1999	Ford acquires Th!nk (formerly PIV/CityBee)	ZEV legislation in California requires selling zero-emission vehicles, pressures automobile-production regime to change. Th!nk had been nurtured in hydropower and public transportation regime, but acquired by automobile regime actors	Reduced annual license (1996), road-toll exemption (1997), reduced taxable benefit on company cars (1998)	Norway: public service, company fleets, and some private customers
2003	Ford sells Th!nk	Chrysler and General Motors win lawsuit against California, ZEV becomes less strict. Th!nk is now without incumbent automobile regime actor support	VAT (25%) exemption (2001), local experiments with bus lane access (2003)	3000 vehicles sold in Norway
2004	Th!nk goes bankrupt	Company unable to subsist in automobile regime without incumbent support		Small, private, urban market. Mainly import
2009	Car dealers begin importing EVs for mass market	Climate change as landscape is developed. Policies earlier intended to stimulate industry development now helps Norwegian vehicle market EV transition	Bus lane access permanent (2005), ferry ticket exemption (2009)	3347 EVs registered (2010)
2013	EV market takes off	Positive user experiences produce new narratives about EVs. Positive media attention		19,678 EVs registered. 500 chargeable hybrids registered
2017	The sale of new EVs higher than the sale of new fossil cars	Large automobile regime actors use Norway as test bed for new models. Alternative regime actors (Tesla) have Norway as key market. Some public controversy on EV incentives		126,448 EVs, 58,213 chargeable hybrids registered

8.4 Conclusion

Following the multi-level logic, a transition policy should follow a two-way strategy: (a) Stimulate the emergence and diffusion of niche innovations, and (b) enhance selection pressure on the regime through economic instruments (such as carbon taxes) and regulation (Geels 2012). Although transportation policies pay moderate attention to the first strategy, and little attention to the second one (Geels 2012), it can be easy to conclude that the current success of electro-mobility in Norway was produced by the second strategy. Such a shortcut would grossly oversimplify the narrative and lead to neglecting the industrial ambitions that once underpinned the development of this (policy) strategy. Looking back, it is difficult to say how successful Norwegian policies for stimulating demand would have been without these industrial ambitions or if the strategy would even have emerged without its industrial predecessor. Furthermore, focusing too much on the effects of policy on technology development can lead to neglecting the political processes that bring about policy change (Normann 2015). There has been an increased focus in recent years on forming policy (e.g., Weber and Rohracher 2012; Normann and Hanson 2017; Kern and Rogge 2017) when studying socio-technical transitions. Our chapter contributes to this growing body of scholarship.

From the literature on socio-technical transitions, we know that niches are often sustained through demonstrations or experimental projects, which allow niche actors to learn about innovations in real-life circumstances. Niches tend to gain momentum if visions and expectations "be-come more precise and more broadly accepted, if the alignment of various learning processes results in a stable configuration ('dominant design'), and if social networks become bigger (especially the participation of powerful actors may add legitimacy and bring more resources in-to niches)" (Geels 2012: 4). This resonates well with the Norwegian case, in which EV driving was initially pioneered by actors that were not involved in the automobility regime. They could act in this capacity because big car manufacturers had not yet moved into these areas. When they did, they often created strategic alliances with small firms or took them over (Dyerson and Pilkington 2005). Our analysis also highlights that alliances with dominant regimes might be treacherous, as the shielding, protection, and flexibility of being a niche actor might be lost as a result.

As shown in the previous sections, the industrial strategy to develop EVs in Norway contributed to developing many of the incentives that we find today. However, it seems non-intuitive that strong policies were related to attempts to nurture a niche for EV production as an alternative to combustion-engine developments. The first serious efforts to commercialize Norwegian EVs (Th!nk) were launched in the late 1990s, out of the desire to establish a Norwegian EV production. However, local air quality, energy efficiency, and increased use of Norwegian electricity were also important ingredients in the work to establish EVs as a promising technology. Environmental organizations worked toward creating favorable user conditions for EVs and demonstrating the assets of EVs contributing to many current local and national EV incentives:

> [T]he incentives have been added one at the time until the market finally responded with in-creased sales… the prolonged EV interest and lobby organizations that fought for better incentives is what resulted in Norway having the largest EV incentives in the world the largest EV fleet and yearly sales per capita. (Figenbaum and Kolbenstvedt 2013)

Our analysis demonstrates that this narrative is too simplistic. We must go many years back in time to understand the Norwegian attempts to develop an alternative car manufacturing industry to better understand why Norway chose to introduce these incentives. When observed as a longitudinal process, it becomes clear that landscape changes and external events (such as the oil crisis and the sudden change in Californian policies) have been essential to the trajectory of Norwegian EV developments. Furthermore, when Norwegian niche actors tried to enter international automobility regime, they became vulnerable to changes, volatilities, and fluctuations, from which they were previously shielded. This study contributes to the transition literature by highlighting how national niches sometimes depend on international regimes for support, but that the actions in these very regimes might sometimes destabilize local niches.

While the MLP has a strong temporal orientation, the spatial dimension has been less elaborated (Geels 2012; Raven et al. 2012). The complications this creates for the transportation domain are clearly visible in this analysis, since many dimensions of the automobility regime are national or international, while some are local. This results in the fact that national mobility regimes can have local variations, and local actors may also support more radical niche projects that can form the seeds for future transitions (Geels 2012).

There have been calls to elaborate further on the spatial dimension of transitions (see Bulkeley et al. 2010; Geels 2012; Raven et al. 2012). Building on this argument, it is interesting to revisit recent debates on the relationships between technological-innovation systems and space, which focus on how industries located in one country may relate to international technological-innovation systems (Normann and Hanson 2017). A common argument in this debate is that a lack of domestic market also represents a barrier for internationalization. In light of the analysis of the Norwegian attempt to develop a technological niche market of EVs (and the automotive industry's long-term lack of a Norwegian domestic market), the dynamics are even more complex. The work by Norwegian actors to access international markets contributed to making the Norwegian alternative automotive industry more vulnerable, rather than more robust as we might anticipate. Later, the market niche created by comprehensive Norwegian support mechanisms for introducing EVs benefited actors in the international technological-innovation system. Meanwhile, this indicates that market demand can be actively created by active, comprehensive political nurturing, as demand factors are one of the biggest challenges for introducing a new technology. However, the role of EV users was not very significant during the 1990s, when the incentives were introduced. Environmental NGOs represented users in their battles to provide local traffic-related benefits for the few EVs on Norwegian roads.

In sum, the Norwegian EV transition was a two-stage process. The first stage (1990–2009) focused on nurturing a domestic EV industry. During these two decades, a comprehensive package of policies was introduced. However, the actual Norwegian market for EVs remained limited. The second phase (2009 to present) focused less on industry development. Today, the only way to obtain one of the few Norwegian EVs that remains is on the secondhand market, and the dream of a Norwegian EV industry resembles a distant memory. Instead, this period is characterized by changes in practices, discourses, perceptions, and mobility culture. Today, EVs are mainstream, and most EV drivers report that their EVs are better and more comfortable than gas-driven cars (Ingeborgrud and Ryghaug 2017). The emergence of the Norwegian EV culture appears to have been influenced by landscape developments, primarily climate change. The pleasure related to driving green, nonpolluting cars has also been very important, sometimes more so than the economic benefits (Ryghaug and Toftaker 2014).

Norwegian EV incentives are likely to be reduced and changed in the years to come, as the number of electric cars grows. For instance, EV owners must contribute to the costs of maintaining transportation infrastructures in the long run. At the same time, it seems reasonable to expect continued technological development and design of new EV models and more shared mobility solutions, as a result of digitalization of the transportation sector and new platform solutions. Although these practices have contributed to a new discourse of sustainable mobility, Sheller (2012: 191) notes: *It still remains questionable to what extent these cultural shifts will impact on the overwhelmingly automobile-centered pattern of majority mobility.* Recent growth in car sales in Norway, and Norwegians' continued fondness for their privately owned electric cars, also contributes to such an understanding, although national policies and city municipalities are forcefully pushing for limiting the use of cars in inner cities in favor of promoting walking, cycling, and public transportation in new, reinforced ways. The Norwegian case is intriguing, as it is one of very few in which electrification of the transportation sector seems to be well underway. However, we may ask to what extent replacing ICs with EVs really transforms our mobility system. Discussions with key experts in the Norwegian mobility sector (Ryghaug and Toftaker 2016) reveal that individual car use still seems to be the dominant mode of transportation and that they see deployment of technology as first and foremost relying on techno-economic incentives. Alternative trajectories portraying more changes in mobility patterns and culture through more travel planning, mixed use of multiple transportation modes, less private car ownership, car sharing, and more investments in modal transfer and parking spaces to allow for the aligning transportation modes exist but do not seem to be underpinned by dominant imaginaries or policies. Therefore, the EV transition in Norway might not significantly affect the automobile-centered patterns of mobility. There is the risk of potentially entering an electric-automobility system if policies promoting, for example, shared and integrated transportation are not also developed quickly.

References

Aamaas B, Peters GP (2017) The climate impact of Norwegians' travel behavior. Travel Behav Soc 6:10–18

Åm H (2015) The sun also rises in Norway: solar scientists as transition actors. Environ Innov Soc Transit 16:142–153

Asphjell A, Asphjell Ø, Kvisle H (2013) Elbil på Norsk. Transnova, Oslo

Aune M (2007) Energy comes home. Energy Policy 35(11):5457–5465

Bjerkan KY, Nørbech TE, Nordtømme ME (2016) Incentives for promoting battery electric vehicle (BEV) adoption in Norway. Transp Res Part D Transp Environ 43:169–180

Buland T (1994) Framtiden er elektrisk? IFIM-notat 4/94

Bulkeley H, Broto VC, Hodson M, Marvin S (eds) (2010) Cities and low carbon transitions. Routledge, New York

Crettenand N, Finger M (2013) The alignment between institutions and technology in net-work industries. Compet Regul Netw Ind 14(2):106–129

Dyerson R, Pilkington A (2005) Tales of creative destruction and the opportunistic incumbent: the case of electric vehicles in California. Technol Anal Strateg Manag 17(4):391–408

Ellingsen LAW, Singh B, Strømman AH (2016) The size and range effect: lifecycle greenhouse gas emissions of electric vehicles. Environ Res Lett 11:054010

Farla J, Markard J, Raven R, Coenen L (2012) Sustainability transitions in the making: a closer look at actors, strategies and resources. Technol Forecast Soc Chang 79(6):991–998

Figenbaum E, Kolbenstvedt M (2013) Electromobility in Norway—experiences and opportunities with electric vehicles. Institute of Transport Economics Norwegian Center for Transport Research. Oslo. TØI report: 1281. ISBN 978-82-480-1465-2 Electronic version

Figenbaum E, Assum T, Kolbenstvedt M (2015) Electromobility in Norway: experiences and opportunities. Res Transp Econ 50:29–38

Geels FW (2002) Technological transitions as evolutionary reconfiguration processes: a multi-level perspective and a case-study. Res Policy 31(8):1257–1274

Geels FW (2010) Ontologies, socio-technical transitions (to sustainability), and the multi-level perspective. Res policy 39(4):495–510

Geels FW (2012) A socio-technical analysis of low-carbon transitions: introducing the multi-level perspective into transport studies. J Transp Geogr 24:471–482

Geels FW, Schot J (2007) Typology of sociotechnical transition pathways. Res Policy 36(3): 399–417

Geels FW, Tyfield D, Urry J (2014) Regime resistance against low-carbon transitions: introducing politics and power into the multi-level perspective. Theory Cult Soc 31(5):21–40

Giddens A (1984) The constitution of society: outline of the theory of structuration. University of California Press

Gjøen H, Hård M (2002) Cultural politics in action: developing user scripts in relation to the electric vehicle. Sci Technol Human Values 27(2):262–281

Hård M, Jamison A (1997) Alternative cars: the contrasting stories of steam and diesel automotive engines. Technol Soc 19(2):145–160

Hawkins TR, Singh B, Majeau-Bettez G, Strømman AH (2013) Comparative environmental life cycle assessment of conventional and electric vehicles. J Ind Ecol 17(1):53–64

Hoogma R, Kemp R, Schot J, Truffer B (2002) Experimenting for sustainable transport. The approach of strategic niche management. Spon Press, London

Hopkins D, Stephenson J (2014) Generation Y mobilities through the lens of energy cultures: a preliminary exploration of mobility cultures. J Transp Geogr 38:88–91

Ingeborgrud L, Ryghaug M (2017) User perceptions of EVs and the role of EVs in the transition to low-carbon mobility. In: ECEEE Summer Study Proceedings, pp 893–900

Kern F, Rogge KS (2017) Harnessing theories of the policy process for analysing the poli-tics of sustainability transitions: a critical survey. Environ Innov Soc Transit. https://doi.org/10.1016/j.eist.2017.11.001

Lorentzen E, Haugneland P, Bu C, Hauge E (2017) Charging infrastructure experiences in Norway-the world's most advanced EV market. In: EVS30 symposium. Stuttgart, Germany, 9–11 Oct 2017
Normann HE (2015) The role of politics in sustainable transitions: the rise and decline of offshore wind in Norway. Environ Innov Soc Transit 15:180–193
Normann HE, Hanson J (2017) The role of domestic markets in international technological innovation systems. Ind Innov, 1–23
Norwegian EV owners survey (2017) The Norwegian EV owners survey 2017. http://www.elbil.no/om-elbilisten-2017. Accessed June 2017
Østby P (2004) Educating the Norwegian nation: traffic engineering and technological diffusion. Comp Technol Transf Soc 2(3):247–272
Raven R, Schot J, Berkhout F (2012) Space and scale in socio-technical transitions. Environ Innov Soc Transit 4:63–78
Røste R (2001) Næringspolitikk for konkurransedyktige nyetableringer-en casestudie av den elektriske bilen Think fra idé til marked. M.Sc. thesis, University of Oslo
Røste R (2013) Value chain analysis of the Norwegian electric vehicles market—think a first-mover. Nifu rapport
Ryghaug M, Toftaker M (2014) A transformative practice? Meaning, competence, and material aspects of driving electric cars in Norway. Nat Cult 9(2):146–163
Ryghaug M, Toftaker M (2016) Creating transitions to electric road transport in Norway: the role of user imaginaries. Energy Res Soc Sci, 119–126
Scott WR (1995) Institutions and organizations. Sage, Thousand Oaks
Sheller M (2014) The new mobilities paradigm for a live sociology. Current Sociol 62(6):789–811
Sheller M (2012) Emergence of new cultures of mobility: stability, openings, and prospects. In: Geels FW, Kemp R, Dudley G, Lyons G (eds) Automobility in transition? A socio-technical analysis of sustainable transport. Routledge, London
Skjølsvold TM, Ryghaug M, Dugstad J (2013) Building on Norway's energy goldmine: policies for expertise, export, and market efficiencies. Renew Energy Gov, 337–349
Smith A, Raven R (2012) What is protective space? Reconsidering niches in transitions to sustainability. Res Policy 41(6):1025–1036
Throndsen W, Skjølsvold TM, Ryghaug M, Christensen TH (2017) From consumer to prosumer: enrolling users into a Norwegian PV pilot. In: ECEEE summer study proceedings
Weber KM, Rohracher H (2012) Legitimizing research, technology and innovation policies for transformative change: combining insights from innovation systems and multi-level perspective in a comprehensive 'failures' framework. Res Policy 41(6):1037–1047

Marianne Ryghaug holds a Ph.D. in Political Science and is a Full Professor of Science and Technology Studies at the Norwegian University of Science and Technology. She is Deputy Director of the Centre for Sustainable Energy Studies (CenSES), a national center for environment-friendly energy research. She has been engaged in energy- and climate-related research since 1999 and has published widely on these topics in top international journals. Her areas of expertise include energy and climate policy, sustainability transitions and innovation policy, and studies of users, practices, and public engagement. Her recent research has focused particularly on the development, implementation, and use of smart grids, smart homes, and electric vehicles.

Tomas Moe Skjølsvold is a sociologist with a Ph.D. in Science and Technology Studies who works as an Associate Professor at the Norwegian University of Science and Technology. His current research focuses on energy transitions, mobility and societal change through processes unfolding at the intersection of innovation, techno-scientific practice, public engagement, and participation. He has published extensively on such issues over the last years and is currently Editor of the Nordic Journal of Science and Technology Studies.

Part IV
Governance of Integrated Mobility

Chapter 9
Governing Mobility-as-a-Service: Insights from Sweden and Finland

Göran Smith, Steven Sarasini, I. C. MariAnne Karlsson, Dalia Mukhtar-Landgren and Jana Sochor

Abstract Based on a review of recent developments in Sweden and Finland, this chapter analyzes the roles of public organizations in the governance of a transition to Mobility-as-a-Service (MaaS). In particular, we draw on insights from transition frameworks to explore what these two pioneering cases can teach us about how the public sector can both enable the development of MaaS and steer the development trajectory toward diffusion of MaaS offerings that contribute to transport policy goals. We propose three main points. Firstly, public sector organizations at national, regional, and local levels have key roles to play in potential transitions to MaaS, regardless of their intended operative roles in the emerging MaaS ecosystem. Secondly, a central task for public sector organizations is to align operational and tactical MaaS governance activities with both an overarching MaaS strategy and with other relevant strategies, such as transport infrastructures investments, programs for economic and industrial growth, city plans, and parking norms. Thirdly, new models and tools for public–private collaboration are needed in order to effectively govern the development and diffusion of sustainable MaaS.

G. Smith (✉) · I. C. M. Karlsson · J. Sochor
Chalmers University of Technology, Gothenburg, Sweden
e-mail: goran.smith@chalmers.se

I. C. M. Karlsson
e-mail: mak@chalmers.se

J. Sochor
e-mail: jana.sochor@ri.se

G. Smith · D. Mukhtar-Landgren
K2—The Swedish Knowledge Centre for Public Transport, Lund, Sweden
e-mail: dalia.mukhtar-landgren@svet.lu.se

G. Smith
Västra Götalandsregionen, Gothenburg, Sweden

S. Sarasini · J. Sochor
RISE Viktoria, Gothenburg, Sweden
e-mail: steven.sarasini@ri.se

D. Mukhtar-Landgren
Lund University, Lund, Sweden

M. Finger and M. Audouin (eds.), *The Governance of Smart Transportation Systems*, The Urban Book Series, https://doi.org/10.1007/978-3-319-96526-0_9

Keywords Mobility-as-a-Service · Transport policy · Public governance
Sustainable transitions

9.1 Introduction

Mobility-as-a-Service (MaaS) has received a great deal of attention from both researchers and practitioners in recent years. The concept has been proposed, and in limited cases proven, to address impediments to the multimodal use of servitized transport modalities, such as the hassles of planning, booking, and paying, as is the case when transport services are offered by different organizations and through different end user interfaces. It has also been argued that the diffusion of MaaS may completely change both how we travel and how personal transportation is organized and that MaaS could be an emerging trillion-dollar industry at the expense of the incumbent private car sector.

Consequently, a set of underlying objectives for enabling MaaS has been suggested. Firstly, MaaS is proposed to have the potential to address the negative externalities of personal transport by reducing private car use (e.g., Sarasini et al. 2017; Sochor et al. 2015) in order to combat issues such as congestion, parking, noise, transport-related injuries and deaths, local pollution, and carbon gas emissions. This objective is most prevalent in urban and suburban areas (e.g., Aapaoja et al. 2017). Secondly, MaaS is suggested to be able to contribute to increased accessibility to personal transport services by improving and extending the transport service ecosystem (e.g., Melis et al. 2017). For instance, MaaS could complement traditional public transport by offering more agile solutions for rural dwellers. Thirdly, MaaS might increase the efficiency of public spending on transport by facilitating the use of private services that better fit with tasks such as special needs transport (e.g., Heikkilä 2014). Fourthly, MaaS could contribute to the economy by creating space for new innovations and private businesses within the personal transport sector (ibid.).

In coining the term, Heikkilä (2014) described MaaS as "a system, in which a comprehensive range of mobility services are provided to customers by mobility operators" (p. 8). While several definitions of MaaS have been offered since, none has become the de facto standard. For the purposes of this chapter, we understand MaaS as an integrative concept that bundles different transport modalities into joint, seamless service offerings, as a means of providing tailored mobility solutions that cater for end users' travel needs (Mukhtar-Landgren et al. 2016).

Numerous MaaS-related pilot programs have been performed, including Smile in Austria; Qixxit, Moovel, Switchh, and Hannovermobil in Germany; Whim in Finland; and UbiGo in Sweden. While these have often reported promising results in terms of promoting more sustainable transport behaviors (e.g., Sochor et al. 2016), successful transformations from pilots to large-scale implementations are yet to appear. Several innovation barriers have been found to hinder such transformations. While some research has focused on technical impediments—especially the lack of open, interoperable, and trusted interfaces for data (e.g., Li and Voege 2017)—most attention has, arguably, been on organizational and relational challenges. MaaS builds on the integration of offerings from several transport service providers. Thus, MaaS is an intrinsically collaborative venture that requires new business ecosystems to emerge in order to bring MaaS offerings to end users (Smith et al. 2017a). Business models that are viable for all the organizations in the emerging MaaS ecosystems, and that cater for MaaS offerings that contribute to policy goals, are yet to be proven (Sarasini et al. 2017). Similarly, a lack of suitable processes for, and experience in, managing collaborative innovation has been found to make it difficult for key stakeholders to agree on shared goals for MaaS and to divide responsibilities (Smith et al. 2018).

To date, Sweden and Finland have acted as pioneers in the development of MaaS. For instance, the 2014 pilot of UbiGo in Gothenburg (SE) is often referred to as the first demonstration in real-life conditions (e.g., Sochor et al. 2015), while the 2016 launch of Whim in Helsinki (FI) drew international attention to the concept. In both countries, the public sector has had a hand in MaaS developments. The public transport authority (PTA) in the region of Västra Götaland (SE) attempted to procure MaaS (Smith et al. 2017b), and the Finnish Ministry for Transport and Communication (Liikenne- ja viestintäministeriö, LVM) has been praised for its reform of transport legislation, partly motivated by the desire to enable MaaS (Smith et al. 2017c). Nevertheless, public organizations in both countries are still struggling to identify their roles in enabling and governing the development of MaaS (Mukhtar-Landgren and Smith 2018). On one hand, too much regulation might impede the private sector's ability to participate and innovate, leading to unattractive MaaS. On the other hand, too little regulation might lead to MaaS that does not serve public interest (Smith et al. 2017a). Moreover, governing MaaS developments is a complex challenge given that the concept is proposed to challenge prevalent private car ownership. 'Automobility' is deeply entrenched in terms of institutionalized structures on both individual and societal levels (such as lifestyles, markets, and legislation). Thus, holistic, collaborative approaches to governance are presumably needed for the efficient development and diffusion of sustainable MaaS.

Driven by both empirical relevance and research interest, we utilize Sweden and Finland as empirical cases in order to explore how public organizations can govern MaaS in the early stages of its development, both to enable disruptive innovations and to steer the development trajectory toward the diffusion of MaaS offerings that contribute to the fulfillment of transport policy goals. Inspired by Sarasini and Linder (2017), we probe the governance challenge by drawing insights from the literature on transition management (e.g., Kemp et al. 2007; Loorbach 2010), focusing on the role of the public sector in governance activities. In particular, our study addresses the following research question: *How can public organizations create institutional arrangements that are conducive to the development and diffusion of sustainable MaaS?*

Our analysis synthesizes the findings reported in four previous conference papers (Mukhtar-Landgren and Smith 2018; Smith et al. 2017a, b, c). The primary data sources are recorded and transcribed semi-structured interviews with 31 representatives of key public and private MaaS stakeholders, extensive participatory observation, and a structured review of related policy documents.

The chapter is divided into six sections, of which this is the first. Next, the second section introduces transition management. The third section describes how MaaS has developed so far in Sweden and Finland. The fourth section outlines what types of governance activities have been undertaken by different types of Swedish and Finnish public organizations in relation to the development of MaaS. The fifth section analyzes the approaches to governance in the two cases and proposes takeaways for public sector organizations. Finally, the sixth section suggests potential topics for future transition-oriented research on MaaS.

9.2 Transition Management

Transition management (TM) is one of several transition frameworks concerned with the governance of systemic transformations of sociotechnical systems, usually with sustainability as the overarching goal (Markard et al. 2012). TM acknowledges the potential roles of multiple stakeholders from different societal sectors (government, industry, research, consultancy, civil society, grassroots movements, etc.) in sustainable transitions (Loorbach 2010). This approach mirrors developments in the wider field of environmental governance, where the term has been broadened to acknowledge the role of non-state organizations in governance activities (Driessen et al. 2012). Traditionally, governance has been seen as synonymous with the conditions upon which public policies are framed and acquire content following interactions between ensembles of organizations in a given institutional context (Kickert et al. 1997). By contrast, the 'government-to-governance shift' (Hysing 2009) has served to redefine governance in terms of multi-stakeholder involvement (Glasbergen 1998), following the empowerment of civil society organizations and the rise of private sector self-governance activities. TM reflects this shift by proposing a prescriptive, collaborative, and multi-stakeholder governance program that relies on co-creation and social learning (Kemp et al. 2007).

While recognizing the importance of multi-stakeholder engagement in the governance of transitions, we delimit our case to an examination of the roles of public sector organizations for two principal reasons. Firstly, our analysis focuses on the role of institutional arrangements in enabling and hindering MaaS developments. Although institutions are a broad concept, we posit that the public sector plays an important part in defining and orchestrating institutional conditions generally, especially in Scandinavian countries, which have been described as 'coordinated' market economies due to tight links between industry and the state (Soskice and Hall 2001). Secondly, MaaS is commonly described in these countries as an innovative concept whereby public transport constitutes the backbone of combined services (e.g., Holmberg et al. 2016). This entails that MaaS developments largely rest on the willingness of public sector organizations to create a set of conditions that enable new innovations to emerge in the MaaS field. In practice, existing public transport systems are maintained by a set of entrenched institutional arrangements (such as procurement rules, ticketing schemes, buses and trams, routes and timetables, and organizational cultures). Hence, while MaaS developments are contingent upon transforming the structures that maintain systems of 'automobility' (that is, private car ownership and use), they are also influenced by the institutional arrangements associated with the public transport system. Others have noted the complexities of this transition, with MaaS described as being 'caught between two regimes', namely those related to public transport and private car ownership or 'automobility' (cf. Parkhurst et al. 2012). The existence of these two regimes makes the governance of MaaS developments a challenging task.

The concept of a 'regime' is drawn from another transition framework—the multi-level perspective (MLP) (Geels 2002). TM studies often utilize the MLP and particularly the concept of a sociotechnical regime (Rip and Kemp 1998) to identify drivers and barriers of system innovations. By using the MLP in this way, as a heuristic device, TM scholars outline governance implications based on a detailed understanding of dynamics of system innovation. Within the MLP, sociotechnical regimes are viewed as a major source of stability, inertia, and lock-in effects, which makes them, arguably, the source of many of the barriers to sustainable transitions. Regimes are multi-actor networks in which the propensity for regime organizations to utilize existing heuristics results predominantly in incremental rather than radical innovation. The structuring qualities of regimes come from numerous sources. Firstly, organizations are embedded within a system of institutional arrangements that enables and hinders certain activities (Geels 2004). Secondly, the organizations within regimes are bound by interdependencies between organizations and networks (Geels 2002). Thirdly, artifacts and material elements of regimes acquire certain durability over time. The artifactual elements of large technical systems, such as electricity infrastructures, acquire 'a logic of their own' due to complementarities with other system elements and sunk costs (Rycroft and Kash 2002). The regime related to private car ownership is embedded in a multilayered institutional context that contains various regulations, norms, and cultural understandings, and also relies upon different types of physical infrastructure, markets, and the car as an artifact per se (Urry 2004).

To overcome regime inertia, TM acknowledges the importance of cycles of learning and adaptation throughout the innovation process, which is commonly divided into four phases: pre-development, takeoff, acceleration, and stabilization (Nevens et al. 2013). TM also recommends a long-term approach that is intended to overcome the short-termism associated with political cycles and the private sector, consisting of four iterative steps: (i) *strategic* (envisioning) activities, which focus on the creation of adaptable, long-term visions that are created by and embedded among relevant organizations; (ii) *tactical* activities, which link individual organization strategies to shared long-term visions; (iii) *operational* activities, which link everyday activities and innovative experiments to long-term visions and can focus on experiments with new products/services, new policies and legislation, and social innovations; and (iv) *reflexive* activities which focus on iterative monitoring, assessment, and evaluation of experimental policies and practices as a means of revising overarching visions and plans where necessary (Kemp et al. 2007). Hence, one way to understand why transitions occur in a particular place (and not in others) is to evaluate the manner in which these activities are being practiced.

One critique of transition frameworks such as the MLP is that it obscures the role of spatial scales (that is, geographical conditions) in transitions, focusing instead on temporal and structural variables (Raven et al. 2012). This critique is based on the premise that the MLP is an adaptive framework that allows researchers to willfully delineate system boundaries, with most studies consequently focusing on national settings when applying the regime concept (Coenen et al. 2012). One might be tempted to mirror this approach by treating countries as institutionally homogenous entities whereby MaaS developments unfold. However, the very basis of this critique is that countries are not institutionally homogenous; they are multi-scalar entities within which organizations operate on a local scale, often with supranational influence (Hansen and Coenen 2015). Alternatively, one may suggest that cities are a useful spatial scale, given that MaaS may, initially at least, target urban and suburban citizens. Indeed, some work has been done to apply the tenets of TM to urban settings using terms such as urban transition laboratories (Nevens et al. 2013). Still, while cities may be essential in governing transitions—for instance, by creating niches for experimentation—scholars have also noted that cities do not act alone in seeking to transform regimes, and those that succeed have ties with national governments and other supranational entities (Hodson and Marvin 2010). Such relational ties are key to establishing a set of institutional arrangements that are conducive to the development and diffusion of radical innovations. This distinction is also useful when considering the difference between *absolute* spatial scales (that is, those that are territorial) and *relative* spatial scales (that is, those that are socially constructed), where the latter are seen to be more relevant to sociotechnical transitions (Raven et al. 2012). When attempting to synthesize multi-scalar perspectives into transition frameworks, economic geographers have drawn upon a few useful concepts. Coenen et al. (2012) noted the importance of institutional thickness; that is, "the comparative performance of governance bodies in terms of their ability to work together locally, and persuade or compel sufficient external agents to support their activities" (p. 972). Similarly, Raven et al. (2012) noted that cognitive, organizational, social, and institutional proximity are conducive to innovation. These terms refer to similarities and trust among organizations along different dimensions and can circumvent the need for spatial proximity.

A further critique of TM is that, in addition to targeting sustainable reorientations of sociotechnical systems, it requires major changes in other key areas such as environmental policymaking. In practice, some of the key tenets of TM are 'lost in translation' when transferred from one national context to another, such as between the Netherlands and Finland (Voß et al. 2009). While we acknowledge that these sorts of problems exist, mainly due to prevalent cultures within key sectors and organizations, our aim is not to evaluate the way in which TM, as a *reflexive* policy paradigm, is transferred from one national context to another. Rather, we utilize the TM framework in two ways. Firstly, we evaluate MaaS developments in Sweden and Finland using TM as an analytical framework that can elucidate the key elements required for the governance of a sociotechnical transition. Secondly, we use this framework to elucidate implications for governance. That is, we contend that the public sector can be imperative in creating a set of institutional arrangements conducive to MaaS developments by, intentionally or otherwise, performing *strategic*, *tactical*, *operational*, and *reflexive* governance activities; and ensuring that relational linkages that connect organizations acting at different spatial scales support these activities (local, regional, national, and supranational).

9.3 Developments in Sweden and Finland

Sweden and Finland have arguably been global pioneers in the early days of MaaS. Sweden witnessed the first comprehensive MaaS pilot program in 2013–2014. Over 70 households in Gothenburg trialed UbiGo, a service that bundled public and private transport services to customized packages of digital clip cards. The UbiGo pilot was deemed successful in terms of user acceptance and favoring sustainable travel (e.g., Sochor et al. 2014), but the contract between the regional PTA and UbiGo was not extended after the pilot period as the PTA had to determine what it was legally able, and strategically willing, to do (Smith et al. 2017b). As part of this work, the PTA decided to initiate a pre-commercial procurement process, looking for a private entrepreneur that could develop, deploy, and operate a comprehensive MaaS solution across the region of Västra Götaland.

Concurrently, the notion of MaaS started growing in Finland. In interviews with central organizations in Finland, the notion is described as having been proposed, and named, by the future founder and CEO of MaaS Global in a government-led think tank, then further developed and detailed in an Aalto University-based master's thesis (Heikkilä 2014), and eventually popularized through the 2014 European Congress on Intelligent Transportation Systems in Helsinki. Two different types of governance processes were initiated in Finland during the years that followed. Firstly, several MaaS projects and pilots were funded and implemented between 2015 and 2016: Sonera Reissu, Ylläs Around, Kätevä, Whim, and Tuup. Secondly, LVM adopted MaaS as a vision for the future organization of the Finnish transport system. The first phase of its ongoing major legislative reform, which will be enacted in 2018, is arguably partly designed to facilitate the development and diffusion of MaaS in Finland as well as the export of MaaS-related innovations (Smith et al. 2017c).

The developments in Finland and the increasing international interest inspired further debates in Sweden. The attention to MaaS spread from having initially been concentrated to the region of Västra Götaland, to ultimately including two other urban regions (the county of Stockholm and the Skåne region), as well as national-level organizations. Currently, several new Maas-related pilots are planned in Sweden, and two national development programs have been initiated to further propel the development.

In order to govern MaaS activities in Sweden and Finland appropriately, both in the short term and long term, it is essential for the public sector to understand how institutional arrangements can either encourage or discourage the development and diffusion of MaaS. Institutional drivers and barriers to sustainable transitions, such as the alleged shift to MaaS, can be both formal and informal (Scott 2014) and can arise on different societal levels. The IRIMS framework (Mukhtar-Landgren et al. 2016) delineates institutional arrangements into three analytical levels. The *macro*-level encompasses societal arrangements, such as continental procurement laws and national identities; the *meso*-level includes institutional arrangements at the regional and local levels, such as regional transport directives and local cultures of collaboration; and the *micro*-level reflects the level of the individual (in this case referring to the proposed users of MaaS) and covers the institutional arrangements that impact their behaviors, such as existent transport infrastructures and current travel habits (Karlsson et al. 2017a).

On the *macro*-level, societal trends such as digitalization, servitization, city densification, more flexible work times, higher expectations of positive use experiences, and the growth of the sharing economy are, despite the lack of empirical evidence, often described as general drivers of MaaS (e.g., Tinnilä 2016). Extant research has found that, notwithstanding geographical proximity and similar institutional arrangements, the public sectors' main objectives for enabling MaaS diverge between Sweden and Finland (Mukhtar-Landgren and Smith 2018). In Sweden, public organizations' funding of and involvement in both the UbiGo pilot and the forthcoming MaaS developments can be understood as a response to the identified need to find new cost-effective measures that can contribute to an increased modal share of public transport, which, in turn, can help reduce the negative externalities of personal transport. Also, the public sector in Finland is hoping that MaaS can contribute positive effects on sustainability. However, their interest in MaaS is rather a derivative of its quest to battle economic downturn (Smith et al. 2017c). In terms of *macro*-level barriers, legislation has been proposed to hinder innovation and renewal in the transport sector in general. Correspondingly, continental and national legislations have been found to limit both the Swedish and Finnish PTAs' understandings of what roles they can take in relation to MaaS (Mukhtar-Landgren and Smith 2018), thus constraining their MaaS-related actions.

On the *meso*-level, all organizations are supposedly motivated by MaaS's potential contribution to their organizational goals, whether these goals are increased profit, reduced car traffic, or something else. In Finland, private investments, prosperous cross-sector collaborations, and strong informal networks have been found to further drive the attentiveness to MaaS (Smith et al. 2017c). In contrast, a lack of a shared vision for MaaS as well as few MaaS champions with the discretion and authority to impact high-level decisions seems to have created further challenges in Sweden, compared to Finland (ibid.). The need to identify business models that are viable for all the organizations in the emerging MaaS ecosystems; the transport service providers' unwillingness to open up their tickets for third-party resale; and the lack of data and standards have been identified as key *meso*-level barriers in both countries (ibid.). Further, a case study of a Swedish PTA's MaaS efforts found that its lack of experience of and processes for public–private innovation made it difficult for the PTA to collaborate with private organizations on MaaS developments (Smith et al. 2018). Explicitly, its use of a 'rigid' public procurement processes was pinpointed as hampering experimentation and collaboration, issues that were further augmented by low levels of trust between the PTA and the potential bidders. In similar fashion, the PTA's organizational culture was considered to foster inertia and the PTA struggled with prioritizing MaaS internally (ibid.).

On the *micro*-level, the most prevalent barrier is arguably the limited knowledge about the potential end users. The MaaS development is still in its infancy, and few of the MaaS-related pilots have been systematically evaluated in terms of end users' adoption, use, and the impacts on their travel behaviors (Karlsson et al. 2017b). As a result, both public and private organizations struggle to establish what the potential return of MaaS investments could be. For instance, how many new end users might MaaS attract to public transport within a given geographical area? Further, several institutional arrangements are thought to favor private car use and thus preclude a transition to MaaS. These include existing travel habits, private car lock-in effects, and current taxation rules, such as subsidization of company cars and tax deductions for expenses related to car travel to and from work (Holmberg et al. 2016).

In sum, drivers and barriers on multiple levels affect the development and diffusion of MaaS. The perceptions of these partly differ and partly coincide between Sweden and Finland (Smith et al. 2017c), which implies that the appropriateness of different approaches to governance probably differs somewhat between the countries as well. Distinct differences have also been found in terms of how innovation barriers are perceived among public and private organizations (Smith et al. 2018), suggesting that a shared understanding of what is hindering the development of MaaS is lacking across organizations within the emerging MaaS ecosystems.

9.4 The Roles of the Public Sector

Public organizations in both Sweden and Finland have been actively involved in the developments described above. As the development of MaaS is still in a pre-commercial stage in both countries, the following text focuses on what actions public organizations on national, regional, and local levels have taken in order to (i) ignite the development and (ii) either govern initial MaaS developments or create possibilities to govern the development trajectory for MaaS in the future. Further, drawing on TM, we center the inquiry on *tactical*, *operational*, *strategic*, and *reflexive* governance activities. Our findings are summarized in Table 9.1.

Table 9.1 Summary of identified governance activities

	Sweden	Finland
National authorities	Following regional activities, have stimulated MaaS experimentation and research through *tactical* and *reflexive* activities	Have acted as a spearhead for MaaS developments by promoting a shift to MaaS through *strategic*, *tactical*, and *operational* activities
Regional public transport authorities	Have initiated and assisted MaaS developments through *tactical* and *operational* activities	Following national activities, have pursued *operational* activities that enable MaaS piloting
Local authorities	Have mostly participated in *operational* activities	Have mostly participated in *operational* activities

9.4.1 National Authorities

The national governments in Sweden and Finland have so far taken different roles in the development of MaaS (Mukhtar-Landgren and Smith 2018) and have therefore used dissimilar policy instruments to govern the trajectory of MaaS. While the Swedish government has primarily utilized soft measures, the Finnish government has also used its regulative and legislative powers to initiate change.

In Sweden, MaaS emerged as a regional phenomenon with little involvement from the national government. It was not until late 2016, more than two years after the acclaimed UbiGo pilot, that the Swedish Ministry of Enterprises and Innovation (Näringsdepartementet) took any concrete action in relation to MaaS. Then, the proposed potential of MaaS and the outspoken need for governmental action coincided with Näringsdepartementet's ongoing pursuit to catalyze innovation that might benefit the next generation's travel and transport. An expert group was tasked with drafting a road map for the diffusion of MaaS in Sweden, including how the national government could support the development. This resulted in funding of a development program entitled Combined Mobility-as-a-Service in Sweden (*Kombinerad mobilitet som tjänst i Sverige*, KOMPIS). The program links the plans and actions of several Swedish public organizations. It will run from 2017 to 2020 and distribute approximately two million Euros of public money in order to set the scene for and initiate the diffusion of MaaS in Sweden. The grand vision is that, by 2030, legislation, policies, and transport norms in Sweden should be shifted to favor traveling by 'shared modes'.

In contrast, the development of MaaS in Finland has, from the very start, been tightly coupled to national government activities (Smith et al. 2017c). During the last decade, LVM has been on a quest to transform the Finnish transport sector, aiming to offset the national economic downturn by streamlining public spending on personal transport and by creating space for new digitally driven innovations within the transport sector. Since 2014, LVM has used the enablement of MaaS as both an internal tool for selecting and revising its actions and as a tool for externally communicating its agenda. The ministry's actions have influenced Finnish MaaS developments in two major ways. Firstly, its frequent use of MaaS as a vision for Finland's future transport sector has drawn attention to MaaS from both entrepreneurs and investors. Secondly, its proposed major reform of Finland's transport legislation, the Transport Code, is meant to pave the way for market-driven MaaS (cf. Smith et al. 2017a). The first phase of the reform, which will be enacted in 2018, is meant to make more room for collaborative innovation within transport by deregulating the public transport and taxi markets, and by regulating transport service providers' use of open interfaces. As of July 2018, all providers of road and rail transport services in Finland, including brokering and dispatch organizations, must provide external parties with access to the sales interface of their ticketing and payment systems and allow them to purchase and resell ticket products at a basic price that, at minimum, entitles the end user to a single trip. As such, LVM is actively forcing transport service providers to collaborate with MaaS operators.

The national innovation funding agencies in the two countries have had comparable roles in that they have mainly funded MaaS-related experimentation and research. The Swedish Innovation Agency (Vinnova) funded the research project that included the UbiGo pilot (Go:Smart). It has since funded several other projects that might benefit the development of MaaS, and is also involved in distributing the funding linked to KOMPIS. Further, Vinnova is setting up a 'policy innovation laboratory'. The laboratory is meant to enable institutional experimentation, for instance, regarding the consequences of altering innovation-related regulation, and one of the proposed applications for this is MaaS. In Finland, the Funding Agency for Innovation (Tekes) has funded numerous MaaS-related pre-studies and pilots as part of a joint development program for MaaS, which it has managed together with LVM since 2014.

A handful of other national authorities are participating in MaaS developments. In Sweden, the Swedish Transport Administration (Trafikverket) has included the KOMPIS road map in its proposed 2018–2029 action plan for the national transport system, and the Swedish Energy Agency (Energimyndigheten) is hosting an innovation contest meant to boost the diffusion of MaaS in Sweden. In Finland, the Finnish Transport Agency (Liikennevirasto) and the Ministry for Agriculture and Forestry (Maa- ja metsätalousministeriö) have both funded MaaS pilots as well, while the Transport Agency is also participating in several MaaS-related initiatives in order to better understand its role in MaaS (e.g., the MaaS Alliance and the research project MAASiFiE). Moreover, Finpro, a publicly owned organization, has initiated a growth program for MaaS aimed at challenging Finnish companies to act upon the global business opportunities of MaaS, and at increasing the awareness of Finland as a great investment target for Maas-related innovation work. These activities show that national agencies beyond the transport policy field are also involved in the development of MaaS, mainly those dealing with innovation policies.

In sum, the clearest similarity between the roles of the Swedish and Finnish national authorities is that both are focusing on enabling private entrepreneurship; that is, both countries seem to envision private organizations as MaaS operators (Smith et al. 2017c). They are attempting to realize this through *tactical* governance activities including collaborative governance, that is, organization of formal and informal collaborative networks, and funding of research and development. However, the two nations differ in that the national government in Finland, to this day, has been a superior promoter. It has placed greater effort into *strategic* activities, has been more actively involved (*operational*) in the development of MaaS, and has used 'harder' policy instruments to boost its developments, compared to its Swedish counterpart (Mukhtar-Landgren and Smith 2018). In Sweden, the national authorities have been more reactive and seem to put greater emphasis on *reflexive* governance activities, funding more assessment-oriented research, compared to Finnish authorities, which have been more concerned with boosting development.

9.4.2 Regional Public Transport Authorities

Public transport has frequently been described as the backbone (main transport mode) of MaaS in Europe. As a consequence, PTAs seem to be in a key position to enable the development of MaaS. As of now, they can single-handedly either empower or block MaaS developments by deciding what public transport tickets MaaS operators can purchase and resell, what business deal the MaaS operators get (with the PTAs), and what resale rules they need to adhere to. However, the Transport Code will erase some of this power for Finnish PTAs. In preparation for this change, the PTA of Greater Helsinki (Helsingin seudun liikenne, HSL) decided to author a generic MaaS contract, which enabled it to negotiate an operative contract with MaaS Global and thereby support the widely discussed Whim pilot in Helsinki.

In Sweden, the operative company of the PTA in the region of Västra Götaland, Västtrafik, participated in the UbiGo pilot. It since tried to procure MaaS as a means of progressing from pilot to implementation (Smith et al. 2017b), but discovered that its proposed contract terms did not allow potential bidders (MaaS operators) to develop business models that would be both viable and efficient in contributing to public transport growth, which was (and is) Västtrafik's chief aim in relation to MaaS. The PTA in Stockholm County (Stockholms läns landsting, SLL) entered the MaaS scene at a later stage. Drawing on Västtrafik's alleged missteps, it decided on a MaaS strategy in which it will not procure MaaS, but will instead initiate and participate in pilots in preparation for opening up its tickets for third-party resale, a direction in which Västtrafik also seems to be headed. Moreover, Västtrafik, SLL, and several other regional PTAs in Sweden have joined forces in a development program—the Swedish Mobility Program (SMP)—for MaaS hosted by their joint development company, Samtrafiken. Beyond initiating and coordinating Swedish

MaaS pilots (and other MaaS efforts), the major goals of the program, which was initiated in 2016, are to develop a national integration platform for transport-related services and to establish Samtrafiken as a national MaaS integrator.[1]

In sum, PTAs in both Sweden and Finland are rethinking their responsibilities in light of the potential paradigm shift that MaaS entails; that is, what roles should they possess in a future MaaS ecosystem? In doing so, the potential trade-off between the level of openness and perceived control seems to be the hardest nut to crack. Moreover, there appears to exist a potential conflict between public organizations on the national level—who are keen to revolutionize the transport sector and fulfill visionary targets such as replacing the private car as the go-to solution for mobility (e.g., LVM in Finland and Vinnova in Sweden)—and regional PTAs, who are more focused on improving the existent regime and fulfilling incremental growth goals (Mukhtar-Landgren and Smith 2018). The most evident difference between Sweden and Finland is that the PTAs in Sweden have been more outspoken about their desire to enable MaaS. Further, the Swedish PTAs have been highly involved in creating the 'common' road map for MaaS, while their Finnish counterparts have been largely left out of this process (Smith et al. 2017c). Hence, the Swedish PTAs have arguably actively participated in both *tactical* (such as plans for MaaS and PT growth) and *operational* governance activities (e.g., technical developments, procurement attempts, and pilot participation), while the Finnish PTAs primarily have been involved in the latter.

9.4.3 Local Authorities

Not many local authorities have had 'front-seat' roles in the development of MaaS in Sweden and Finland, with a few notable exceptions. Civil servants at the city of Helsinki have been part of the informal inner circle of MaaS since it first came about in Finland. For instance, the city of Helsinki co-sponsored the master's thesis that was eventually used to promote the MaaS concept during the European Congress on Intelligent Transportation Systems in Helsinki, and has contributed to the drafting of the Transport Code. Moreover, the city of Turku has collaborated with the MaaS start-up Kyyti (formerly known as Tuup), and the city of Tampere has recently launched a three-year MaaS pilot in which it will develop a MaaS platform for the city and trial concepts that make use of spare capacity in special transport services, among other things. In Sweden, the city of Gothenburg was a crucial participant in the UbiGo pilot, but has since become silent, awaiting the actions of Västtrafik, while the city of Stockholm and Lund municipality are planning to participate in MaaS pilots. Still, in both the Swedish and Finnish MaaS developments, the city planner perspective has rarely been discussed; that is, how MaaS might interplay with long-term visions for urban, suburban, and rural developments.

[1]However, these plans have been put on hold for the time being, for financial reasons.

In sum, some local authorities in both Sweden and Finland have been active in the development of MaaS. However, few have contributed to or performed any *strategic* or *tactical* governance activities, and those that have participated have not focused on what MaaS might contribute to, or need from, the physical planning of cities and regions. Instead, they have acted as *operational* enablers of experimentation and have not focused on the long-term vision for MaaS.

9.5 Analysis

The two cases described in this chapter generate useful insights regarding the governance of MaaS developments. Our narrative may be recapitulated as follows. In Sweden, the early success of a thoroughly and well-evaluated pilot was followed by a period of indecision and inaction, in which a superficial understanding may allude to the public sector and particularly the PTA of the region of Västra Götaland having halted MaaS developments. By considering the four key elements of TM, a more nuanced understanding of MaaS developments can be gleaned. Despite making initial headway in terms of *operational* and *reflexive* governance activities, Swedish MaaS developments have slowed—until recently, at least, when the development of a national road map and the establishment of an innovation program have sought to rekindle MaaS-related action. The public sector has played a key role in the attempt to rejuvenate MaaS developments, since the Swedish government succeeded in making MaaS a national priority, supported ably by the research sector and by expert practitioners. The result is a renewed focus on *operational* and *reflexive* governance activities via pilots and evaluations/assessments within the four-year KOMPIS program. Further, by engaging with both PTAs and private transport service providers, and by pushing for a national platform, the SMP has conducted valuable *tactical* activities in cooperation with key operative organizations. Although the plan to establish a national integration platform has been put on hold, these activities have succeeded in stimulating debates within several PTAs and putting MaaS higher on their agendas. The outcome of KOMPIS and SMP, among other activities, appears to be a public–private approach to MaaS in Sweden (cf. Smith et al. 2017a).

The Finnish case, by comparison, when seen through the lens of TM, is in many ways opposite to the Swedish case. The need for economic renewal has forged a stronger national consensus on MaaS, resulting in a more coherent vision (Smith et al. 2017c) and the rapid establishment of an ambitious pilot program, such that the *strategic* and *operational* activities missing in Sweden are prevalent in Finland. Public sector efforts are again prominent. LVM and Tekes have each played significant roles in creating a favorable set of institutional arrangements and supporting research and development efforts with state funding, and a few Finnish municipal governments are active with MaaS debates and developments (although primarily through *operational* activities). One area that appears to be lacking in Finland relates to *tactical* activities that involve the incumbent operative organizations.

Although public and private transport service providers in Finland are supposedly rethinking their roles and positions with regard to MaaS developments, the Finnish market-driven approach (cf. Smith et al. 2017a) appears to require the support of public policies that will coerce them to engage in MaaS ecosystems. For example, the Finnish PTAs have so far been reluctant to participate in MaaS pilots despite the pressure from the national authorities, and the taxi sector has been vocal in its criticism of the reorganizations proposed in the Transport Code.

Both cases imply that geographical perspectives are critical to effective TM activities. By adopting a multi-scalar approach, the importance of relational links between different public sector organizations becomes apparent in terms of *strategic*, *tactical*, and *operational* governance activities. The Finnish case demonstrates the importance of a network of MaaS champions, acting within public and private sector organizations at national, regional, and local societal levels, for creating a robust and legitimate vision for MaaS developments. By comparison, the Swedish case demonstrates the lack of such relational ties as one reason for the apparent slowing of MaaS developments. It also demonstrates the importance of relational ties for *tactical* activities vis-à-vis the SMP, which has arguably enabled a more consensus-based approach to engaging the PTAs. This has helped align key public organizations, albeit at a slower pace than the Finnish Transport Code, which aims to force alignment within public transport organizations. It remains to be seen whether Finnish startups will succeed in developing viable MaaS given the possibility of resistance among PTAs and other transport service providers. In other words, a relational approach to *strategic* and *tactical* activities may be important for aligning interests and engaging organizations such that *operational* activities succeed. Our cases show that public sector organizations at different societal levels can play key roles in facilitating collaborative and relational approaches that can underpin effective TM.

Regarding *operational* activities, the Swedish case demonstrates that traditional mechanisms used by the public sector to drive innovation are unsuitable for collaborative ventures. In particular, conventional public procurement procedures seem to hamper experimentation, encumbering transitions from pilots to implementations and obstructing inter-organizational collaboration (Smith et al. 2018). Moreover, collaborative innovation partnerships between public and private organizations require that the public organizations give up or share some of their governance authority (Bommert 2010). Hence, we propose that, regardless of the operative roles adopted by public organizations, MaaS necessities new models and tools that target a collaborative approach to innovation. In other words, a relational approach is also paramount in *operational* activities.

As noted, the roles of public organizations in Sweden and Finland have diverged, both in terms of what types of governance activities they have performed and what operative roles in the emerging MaaS ecosystem they are planning for. The role of the public sector has been contested in both countries, particularly regarding whether the current roles of PTAs should be broadened or reduced as part of the MaaS transition (cf. Smith et al. 2017a). Which model is most appropriate in this regard remains to be seen. However, we hold firm that pertinent national authorities, regional PTAs, and municipalities should all be involved in drafting and revising an overarching strategy for MaaS, as they, in their capacities as planners of the transportation system, are in key positions to make sure that the potential transition to MaaS is supported by, and in itself supports other public strategies on the national, regional, and local levels. Further, public sector involvement may be a key to the development of a more sustainable transport system.

To conclude, we propose three main points. Firstly, public sector organizations at national, regional, and local levels have key roles to play in potential transitions to MaaS, regardless of their intended operative roles in the emerging MaaS ecosystem. Secondly, a central task for public sector organizations is to align *operational* and *tactical* MaaS governance activities with both an overarching MaaS strategy and with other relevant strategies, such as transport infrastructures investments, programs for economic and industrial growth, city plans, parking norms. Thirdly, new models and tools for public–private collaboration are needed in order to govern the development and diffusion of sustainable MaaS effectively.

Lastly, we revisit our research question: *How can public organizations create institutional arrangements that are conducive to the development and diffusion of sustainable MaaS?* Here, we echo the message from Smith et al. (2017c). Drawing on the analysis presented in this chapter, we argue that the public sector should perform the following, if MaaS is assessed to be an integral part of fulfilling policy goals: (i) engage directive and operative organizations in MaaS networks; (ii) create a strong and shared long-term vision for MaaS that addresses policy goals; (iii) foster an open and collaborative innovation climate around MaaS; (iv) support MaaS pilots and implementations with financial and human capital; (v) experiment with institutional arrangements that could support the development and diffusion of MaaS; and (vi) focus on steering toward societal effects that are positive in the long-term while concentrating on generating and absorbing knowledge in the short-term.

9.6 Concluding Remarks

This chapter is based on two qualitative case studies. This is an appropriate approach for exploring new phenomena in depth and for creating high-quality explanatory theories (Eisenhardt 1989). Still, the generalizability is often questioned (e.g., Miles 1979). Addressing this issue, we side with Donmoyer (1990) on his emphasis on the value of learning from individual cases as well as from reviews of aggregates. In this particular case, we propose a naturalistic approach to the generalizability of our findings; that is, they should be interpreted as transferable to other similar cases rather than to the entire population (Myers 2000). Thus, our proposed insights might be valid for MaaS developments in contexts with similar institutional arrangements as in Sweden and Finland.

Still, as we have traced distinct differences between two neighboring countries, we acknowledge that complementary case studies in dissimilar settings are needed to better establish generalizability. For this reason, we suggest that fellow scholars should further examine the development and diffusion of MaaS in contexts with disparate institutional arrangements, compared to our cases. For instance, studies in countries or regions with dissimilar socioeconomic circumstances, less developed public transport systems, and more autonomous and/or authoritarian governments would complement our study. Further, although both Sweden and Finland may be regarded as global pioneers of MaaS developments, MaaS is nascent in both countries—very few citizens have participated in or experienced any impact from MaaS. Hence, further studies that comprise later stages of the alleged transition, in both similar and dissimilar settings, are needed. Our study has focused on the roles and activities of a limited set of organizations, primarily on the public perspective of governance. Given that MaaS is a collaborative venture, the private perspective should be highlighted as well. Moreover, a few publicly owned organizations that are vital to MaaS have been left out, particularly the national rail companies and organizations on the supranational level. Their roles should be further investigated. Lastly, we recognize that our governance suggestions, while perhaps theoretically sound, remain unproven. Hence, we hope that future research will advance the understanding of how relational, collaborative, and explorative approaches can be materialized.

Acknowledgements This chapter is a product of a research project entitled Institutional Frameworks for Integrated Mobility Services in Future Cities (IRIMS). The authors would like to express gratitude toward Vinnova, which funded the project, and toward our research colleagues who contributed to the analysis, mainly Till Koglin, Annika Kronsell, Adam Laurell, and Emma Lund. Moreover, we are thankful to all the interviewees who devoted their valuable time.

References

Aapaoja A, Eckhardt J, Nykänen L, Sochor J (2017) MaaS service combinations for different geographical areas. Presented at the 24th world congress on intelligent transportation systems, 29 Oct–2 Nov, Montreal

Bommert B (2010) Collaborative innovation in the public sector. Int Public Manag Rev 11(1): 15–33

Coenen L, Benneworth PS, Truffer B (2012) Toward a spatial perspective on sustainability transitions. Res Policy 41(6):968–979. https://doi.org/10.1016/j.respol.2012.02.014

Donmoyer R (1990) Generalizability and the single-case study. In: Eisner EW, Peshkin A (eds) Qualitative inquiry in education: the continuing debate. Teachers College Press, New York, pp 175–200

Driessen PPJ, Dieperink C, Laerhoven F, Runhaar HAC, Vermeulen WJV (2012) Towards a conceptual framework for the study of shifts in modes of environmental governance—experiences from the Netherlands. Environ Policy Gov 22(3):143–160. https://doi.org/10.1002/eet.1580

Eisenhardt KM (1989) Building theories from case study research. Acad Manag Rev 14(4): 532–550

Geels FW (2002) Technological transitions as evolutionary reconfiguration processes: a multi-level perspective and a case-study. Res Policy 31(8):1257–1274. https://doi.org/10.1016/s0048-7333(02)00062-8

Geels FW (2004) From sectoral systems of innovation to socio-technical systems: insights about dynamics and change from sociology and institutional theory. Res Policy 33(6–7):897–920. https://doi.org/10.1016/j.respol.2004.01.015

Glasbergen P (1998) Co-operative environmental governance: public-private agreements as a policy strategy, vol 12. Springer Netherlands, Dordrecht. https://doi.org/10.1007/978-94-011-5143-6

Hansen T, Coenen L (2015) The geography of sustainability transitions: review, synthesis and reflections on an emergent research field. Environ Innov Soc Transit 17:92–109. https://doi.org/10.1016/j.eist.2014.11.001

Heikkilä S (2014) Mobility as a Service—A proposal for action for the Public Administration, Case Helsinki (Master's thesis). Aalto University, Aalto

Hodson M, Marvin S (2010) Can cities shape socio-technical transitions and how would we know if they were? Res Policy 39(4):477–485. https://doi.org/10.1016/j.respol.2010.01.020

Holmberg P-E, Collado M, Sarasini S, Williander M (2016) Mobility as a Service (MaaS): describing the framework. RISE Viktoria, Göteborg. https://www.viktoria.se/projects/maas-framework#publications. Accessed 4 Feb 2017

Hysing E (2009) From government to governance? A comparison of environmental governing in Swedish forestry and transport. Governance 22(4):647–672. https://doi.org/10.1111/j.1468-0491.2009.01457.x

Karlsson ICM, Mukhtar-Landgren D, Lund E, Sarasini S, Smith G, Sochor J, Wendle B (2017a) Mobility-as-a-Service: a tentative framework for analysing institutional conditions. Presented at the 45th European transport conference, Barcelona, 4–6 Oct

Karlsson ICM, Sochor J, Aapaoja A, Eckhardt J, König D (2017b) Deliverable Nr 4—impact assessment. https://research.chalmers.se/publication/248829. Accessed 14 June 2017

Kemp R, Loorbach D, Rotmans J (2007) Transition management as a model for managing processes of co-evolution towards sustainable development. Int J Sustain Dev World Ecol 14 (1):78–91. https://doi.org/10.1080/13504500709469709

Kickert WJ, Klijn E-H, Koppenjan JF (1997) Managing complex networks: strategies for the public sector. Sage Publications, London

Li Y, Voege T (2017) Mobility as a Service (MaaS): challenges of implementation and policy required. J Transp Technol 7(2):e95–e95. https://doi.org/10.4236/jtts.2017.72007

Loorbach D (2010) Transition management for sustainable development: a prescriptive, complexity-based governance framework. Governance 23(1):161–183. https://doi.org/10.1111/j.1468-0491.2009.01471.x

Markard J, Raven RR, Truffer B (2012) Sustainability transitions: an emerging field of research and its prospects. Res Policy 41(6):955–967. https://doi.org/10.1016/j.respol.2012.02.013

Melis A, Mirri S, Prandi C, Prandini M, Salomoni P, Callegati F (2017) Integrating personalized and accessible itineraries in MaaS ecosystems through microservices. Mob Netw Appl, 1–10

Miles MB (1979) Qualitative data as an attractive nuisance: the problem of analysis. Adm Sci Q 24 (4):590–601

Mukhtar-Landgren D, Koglin T, Kronsell A, Lund E, Sarasini S, Smith G, Sochor J (2016) Institutional conditions for integrated mobility services (IMS): towards a framework for analysis. http://www.k2centrum.se/fou-områden/integrerade-mobilitetstjänster. Accessed 4 Sept 2017

Mukhtar-Landgren D, Smith G (2018) Perceived action spaces of public actors in the development of Mobility as a Service. Presented at the 7th transport research arena TRA 2018, Vienna, 16–19 Apr 2018

Myers M (2000) Qualitative research and the generalizability question: standing firm with Proteus. Qual Rep 4(3):9

Nevens F, Frantzeskaki N, Gorissen L, Loorbach D (2013) Urban Transition Labs: Co-creating transformative action for sustainable cities. J Clean Product 50:111–122. https://doi.org/10.1016/j.jclepro.2012.12.001

Parkhurst G, Kemp R, Dijk M, Sherwin H (2012) Intermodal personal mobility: a niche caught between two regimes. In: Kemp R, Geels FW, Dudley G (eds) Automobility in transition? A socio-technical analysis of sustainable transport. Routledge, Oxford

Raven RR, Schot JJ, Berkhout F (2012) Space and scale in socio-technical transitions. Environ Innov Soc Transit 4:63–78. https://doi.org/10.1016/j.eist.2012.08.001

Rip A, Kemp R (1998) Technological change. Battelle Press, Columbus. https://doi.org/10.1007/bf02887432

Rycroft R, Kash D (2002) Path dependence in the innovation of complex technologies. Technol Anal Strateg Manag 14(1):21–35. https://doi.org/10.1080/09537320220125865

Sarasini S, Linder M (2017) Integrating a business model perspective into transition theory: the example of new mobility services. Environ Innov Soc Transit. https://doi.org/10.1016/j.eist.2017.09.004

Sarasini S, Sochor J, Arby H (2017) What characterises a sustainable MaaS business model? Presented at the 1st international conference on Mobility as a Service (ICOMaaS), Tampere, 28–29 Nov

Scott WR (2014) Institutions and organizations: ideas, interests, and identities. Sage Publications, Thousand Oaks

Smith G, Sochor J, Karlsson ICM (2017a) Mobility as a Service: implications for future mainstream public transport. Presented at the international conference series on competition and ownership in land passenger transport (Thredbo15), Stockholm, 13–17 Aug

Smith G, Sochor J, Karlsson ICM (2017b) Procuring Mobility as a Service: exploring dialogues with potential bidders in West Sweden. Presented at the 24th world congress on intelligent transportation systems, Montreal, 29 Oct–2 Nov

Smith G, Sochor J, Sarasini S (2017c) Mobility as a Service: comparing developments in Sweden and Finland. Presented at the 1st international conference on Mobility as a Service (ICoMaaS 2017 Proceedings, pp. 223–239), Tampere, 28–29 Nov

Smith G, Sochor J, Karlsson ICM (2018) Public–private innovation: barriers in the case of Mobility as a Service in West Sweden. Public Manag Rev. https://doi.org/10.1080/14719037.2018.1462399

Sochor J, Karlsson ICM, Strömberg H (2016) Trying out Mobility as a Service: experiences from a field trial and implications for understanding demand. Transp Res Rec J Transp Res Board 2542:57–64. https://doi.org/10.3141/2542-07

Sochor J, Strömberg H, Karlsson ICM (2014) The added value of a new, innovative travel service: insights from the UbiGo field operational test in Gothenburg, Sweden. In: Giaffreda R, Cagáňová D, Li Y, Riggio R, Voisard A (ed) International internet of things summit. Springer, New York, pp 169–175. https://doi.org/10.1007/978-3-319-19743-2

Sochor J, Strömberg H, Karlsson ICM (2015) Implementing mobility as a service: challenges in integrating user, commercial, and societal perspectives. Transp Res Rec J Transp Res Board 2536:1–9. https://doi.org/10.3141/2536-01

Soskice DW, Hall PA (2001) Varieties of capitalism: the institutional foundations of comparative advantage. Oxford University Press, Oxford

Tinnilä MT (2016) Towards servitization of mobility—Mobility as a Service. Int J Res Bus Technol 8(2):958–963. https://doi.org/10.17722/ijrbt.v8i2.444

Urry J (2004) The 'System' of automobility. Theory Cult Soc 21(4–5):25–39. https://doi.org/10.1177/0263276404046059

Voß J-P, Smith A, Grin J (2009) Designing long-term policy: rethinking transition management. Policy Sci 42(4):275–302. https://doi.org/10.1007/s11077-009-9103-5

Göran Smith is an industrial doctoral candidate in innovation for sustainable transitions. In his doctoral project, he studies the development and diffusion of Mobility-as-a-Service (MaaS). The goal is to create better understanding of (i) how institutional arrangements affect disruptive innovations during different phases of their lifecycles, and (ii) how the public sector can enable fruitful innovation collaborations with private firms without losing the ability to ensure societal benefits.

Steven Sarasini holds a Ph.D. in Research Policy from Lund University. He is a Senior Researcher at RISE Viktoria whose research utilizes theories and concepts from transition studies and business model innovation to focus on sustainable forms of MaaS. At present, he is active in projects that examine institutional drivers and barriers to innovation within the MaaS field, action research that aims to stimulate the development of sustainable MaaS services in different urban settings, and assessment of MaaS' sustainability impacts.

I. C. MariAnne Karlsson is a Professor (chair) in Human-technology Systems at Chalmers University of Technology. Her overall research aim is to develop knowledge on the multi-dimensional relationship between people and technology. She has more than 20 years experience of investigating prerequisites for sustainable mobility from a traveler perspective. She has explored the impacts of ICT-based services and has studied MaaS since the birth of the concept.

Dalia Mukhtar-Landgren holds a Ph.D. in Political Science from Lund University, where she is employed as senior lecturer and researcher. She is also a research fellow at K2, the Swedish Knowledge Centre for Public Transport. She has published on topics such as urban development planning and experimental governance. Ongoing research projects include the governing of smart mobility and the usage of new forms of temporary organizations such as pilots and projects in local development and planning.

Jana Sochor holds a Ph.D. in Transport Science from KTH—Royal Institute of Technology. She is a Senior Researcher at Chalmers University of Technology and RISE Viktoria. She has served as a Main Member of the Emerging and Innovative Public Transport and Technologies Committee of the Transportation Research Board since 2016. Her research interests include understanding end users' perceptions, motives, and behaviors as related to MaaS so as to improve service/system design and development, adoption, and effect, as well as positively impact users' mobility choices in a more sustainable manner.

Chapter 10
Institutions, Organizations, and Technological Innovations: The Dynamic Development of Smart Ticketing Schemes in London's Urban Transportation System

Maxime Audouin and Matthias Finger

Abstract The integration of different modes of transportation can be seen as a way of motivating a shift from private car use to more sustainable modes of transportation. Although certain forms of integration have existed for a long time, information and communication technologies (ICTs) have recently opened the way for new possibilities. At the forefront of these developments, smart cards have demonstrated their potential to facilitate multimodal journeys and benefit urban transportation systems. However, while there has been sufficient research into understanding the benefits of smart cards, the present chapter considers the processes that support their emergence. Building on a theoretical framework based on co-evolution theory and systems of innovation, we use archival data to explore the case of the London transportation ticketing system, which has evolved over the last years, from the introduction of the Oyster card in 2004 to the implementation of contactless payment cards, which began in 2012. While it is evident that institutions influence how organizations interact with one another, and how this leads to the emergence of technological innovation, we also highlight the importance of considering the influence that technological innovations have on organizations, which can shape new institutional developments through their interactions. Ultimately, this chapter sheds light on what determines technological innovations in urban transportation systems and, by proposing a new theoretical framework, attempts to bridge the co-evolution and systems of innovation literature.

Keywords Smart transportation cards · Oyster · Contactless payment cards
Co-evolution theory · London

M. Audouin (✉) · M. Finger
Chair Management of Network Industries (MIR), College Management of Technology (CDM), EPFL, Lausanne, Switzerland
e-mail: maxime.audouin@epfl.ch

M. Finger
e-mail: matthias.finger@epfl.ch

M. Finger and M. Audouin (eds.), *The Governance of Smart Transportation Systems*,
The Urban Book Series, https://doi.org/10.1007/978-3-319-96526-0_10

10.1 Introduction

"*Small penis? Have I got a car for you. If you're going to over-compensate, then by all means, over-compensate.*" Those were the words used to advertise the new Porsche 911 Carrera 4 in 1989, summing up the position that the automobile had come to occupy in modern society. Having risen to become the second item of personal consumption after housing, the automobile has become a central element in human existence and is often associated with values of masculinity, speed, professional success, and freedom (Sheller and Urry 2000). This attachment to the automobile is acknowledged to have deeply influenced the form of transportation systems, particularly in cities. Most twentieth century traditional urban planners, who followed the functionalist movement, came to think of cities as being divided into four distinct zones, which were used to work, to live, to enjoy, and to circulate. This approach definitively put automobiles on a pedestal in urban systems, as it made the movements of people between the zones strictly reliant on cars. Modern society progressively became dominated by the automobility concept, defined as a self-organizing path-dependent non-linear system spreading world wide, including cars, infrastructure, drivers, and different technologies, and paving the way for its own self-expansion (Urry 2004). Eventually, automobility, combined with rapid demographic growth, rising motorization, and urbanization, resulted in overcrowded car-centric cities, which ultimately led to detrimental outcomes such as urban congestion and air pollution and their well-known economic, environmental, and social impacts.

Therefore, reducing the number of private vehicles on the road has become a point of fascination for many academics and practitioners, for which two major paths have been proposed. The first path, which might be understood as direct control, deals with the direct internalization of the transport–related costs (such as the creation of a congestion charge). The second path, indirect control, deals with the development of measures aimed at improving the quality and accessibility of non-private motorized transport solutions (Abrate et al. 2009). Among other propositions falling under the second path, multimodal transport—defined by Van Nes (2002: 9) as a combination of "*more than one transport service for making a trip, being combinations of private transport and public transport services or combinations of public transport services*"—has been acknowledged as one possible way of coping with congestion in urban transportation systems. By offering urban dwellers means to reach their final destinations by combining different shared and sustainable transportation modes, multimodal transport has quickly gained traction as a serious alternative to private motorized transport. However, as its name suggests, multimodal transport implies a transfer between different transportation modes, so the development of an integrated transport system becomes paramount for a successful shift from private car to shared and sustainable transport modes (Brons et al. 2009).

Although integrated transport has been defined in several ways (Janic and Reggiani 2001), it can be primarily defined as the pursuit of synergies and the removal of barriers (May and Roberts 1995; May et al. 2005), and it is not a new concept in transport studies. As the first step toward integrated transport policy or integrated transport and land use (Potter and Skinner 2000), modal integration can be divided into different aspects. Physical integration, for example, deals with the design of transportation networks and physical connection points that enable users to move between and within transportation modes. Network integration builds on the existing transportation demand to establish a hierarchy within a given transportation network in order to adapt the system's capacity and decide which modes will be installed for the various transportation branches. Finally, timetable integration involves coordinating the timetables of different transportation modes to ensure quick and easy intermodal transfers.

While these three types of integration have existed for decades, a new form of integration has emerged thanks to the development of information and communication technologies (ICTs). One of the most promising aspects of ICTs has been the associated development of integrated ticketing systems. Also known as smart cards, these ticketing systems support the establishment of both a single fare and a single payment system that is transferable across different transport modes, which frees users of the cash payment burden and reduces boarding times.

Smart cards are considered essential for sustainable urban transport (Goldman and Gorham 2006). Several studies have acknowledged the positive impact of smart cards public transport ridership. For example, Hirsch et al. (2000) demonstrated that between January 1997 and June 1999, the introduction of the MetroCard led to a 20.3% increase in public transport ridership in New York City during weekdays and a 24% increase during weekends. Smart card systems have also been praised for their ability to have positive impacts on the ridership of the other transport modes that are being integrated with public transport. For example, Huwer (2004) demonstrated their benefits for carsharing ridership, while Midgley (2009) reached the same conclusion when looking at the integration of bike-sharing systems with public transport. Furthermore, smart cards have been shown to be cost-effective for transportation authorities. For example, Opurum (2009) showed that the introduction of the New York City MetroCard could save $2.5 million over a 30-year period compared with a do-nothing scenario. Smart cards are also acknowledged to bring other benefits. For example, the T-money card in Seoul has reduced delays caused by cash payments, increased user convenience, and helped ensure the transparency of transportation operators by providing reliable data to the transport authorities (Audouin et al. 2015). Nikolaou et al. (2000) studied the implementation of a smart card system in Thessaloniki, Greece, and showed that smart cards also seem to support a mental shift among car owners, whereby they are increasingly willing to use public transportation rather than their personal cars. Although some research has also been conducted on smart cards from a more technical perspective—for example, by analyzing smart cards' resistance to hackers (Markantonakis et al. 2009)—most research conducted to date has focused on the system's potential economic, social, and environmental benefits.

While there seems to be consensus among scholars that smart cards can improve the conditions and performance of urban transportation systems, very little research has aimed to understand their processes of development or implementation. Exception includes Audouin and Finger (2018). In order to address this gap, the present chapter proposes a theoretical framework, built on the co-evolution and systems of innovation literature, and aims to understand the relationships between institutions, organizations, and technological development. We use archival data to present a case study of the London transportation ticketing system, which has evolved tremendously over the past years, from the introduction of the Oyster card in 2004 to the launch of Contactless Payment Cards in 2014. We then analyze this case in order to understand the dynamic interactions that occurred between institutions and organizations and led to the birth of technological innovations, as well as the impact that these same technological innovations had on organizations and institutions. We finish by discussing the potential paths of the evolution of transport ticketing in London and provide recommendations for further research.

10.2 Theoretical Framework

Albeit a large amount of research has been conducted to look at the development of "hardware" technological innovations in transport (alternative fuels, non-ICE cars), no research, to the best of our knowledge, has investigated the processes underlying the development and implementation of smart transportation cards. The present chapter aims to fill this void. In this section, we propose a theoretical framework built on concepts sourced from literature relating to systems of innovation and the co-evolution of institutions and technology; this framework structures the analysis of our case study.

Over the last few decades, the subject of innovation has gained traction in the social sciences research community (Fagerberg and Verspagen 2009). While some scholars are interested in understanding the impacts and consequences of innovations, the innovations' underlying structure and the conditions that determine their emergence have also garnered significant interest.

At the forefront of this approach lies the systems of innovation school of thought, which aims to understand the "*economic, social, political, organizational and other factors that influence the development diffusion and use of innovations*" (Edquist 1997: 14). Systems of innovation are composed of two main components: first, the organizations, or the players and actors; second, the institutions, which are defined as the rules of the game—these can be formal laws or regulations, common habits, or practices that dictate the relationships between actors (Edquist 2001). The systems of innovation approach was first used at the national level by national innovation systems (NIS) scholars, who described the processes that supported innovation in different countries (see, for example, Nelson (1993) for a comparison of Italian, American, French, British, and Israeli Systems of Innovation). However, NIS was criticized for being overly focused on national boundaries and non-firm organizations and institutions (Malerba 2002). Correspondingly, as innovation is typically bred locally and embedded within a geographic context, scholars proposed regional systems of innovation (RSI) to complement the NIS approach. However, RSI was also criticized, mainly for lacking clarity about whether they were subsets of NIS or strictly new entities (Doloreux 2002). In order to further complement the NIS and RSI approaches, Malerba (2002) proposed sectoral systems of innovation; this approach still built on organizations and institutions, but relied more on individuals and how they interact and influence the system. While sectoral systems of innovation are more able than NSI to conceptualize the impact of institutions on technological innovations (Geels 2004), we believe that they lack perspective, as they do not allow researchers to understand how technological innovations influence the interactions between actors, who can then influence institutional change. While the systems of innovation literature provide a solid framework to understand the determinants of innovation, it fails to describe how these very same innovations inevitably interact with their own determinants. More precisely, in our opinion, it lacks a dynamism that is essential in order to fully understand the relations between institutions, organizations, and technology.

On the other hand, the literature on the co-evolution of institutions and technology has focused on the interactions between institutions and technological development in socio-technical systems and could be a promising source for the dynamic dimension that is lacking in the systems of innovation literature. According to co-evolution theory, the degree to which institutions influence technological development is paralleled by technology's capacity to drive institutional evolution through the interaction of involved stakeholders. This body of literature opens new perspectives that are not clearly defined in the systems of innovation literature. However, we argue that the co-evolution theory lacks clarity on certain points. First, it is not clear how technological development relates to innovation. According to Crettenand and Finger (2013), actors can trigger innovation when institutions and technology are misaligned, but it is not clear how this happens in reality. For Finger et al. (2010), it is the combination of all the actors that create the dynamics that lead to technological or institutional changes, but no further details are given on how these actors concretely interact with one another within and across their categories (institutional, market, and technological).

Thus, we believe that combining these two literatures—systems of innovation and co-evolution between institutions and technology—gets the best of both worlds. Incorporating co-evolution into the systems of innovation theory would introduce a dynamic dimension in which technological innovations can influence organizations and institutions through a feedback loop of sorts. Further, because the co-evolution approach was developed to study the evolution of socio-technical systems, it is also more relevant to urban transportation systems than the systems of innovation approach alone, which was not developed with regard to any particular sector. On the other hand, by also including elements from the systems of innovation approach, we can better distinguish the different actors involved, as well as how their interactions translate into innovation creation. Our theoretical framework is summarized in Fig. 10.1.

Transitions scholars have studied innovation development and implementation, as well as institution-specific investigations into actor-relations, using frameworks

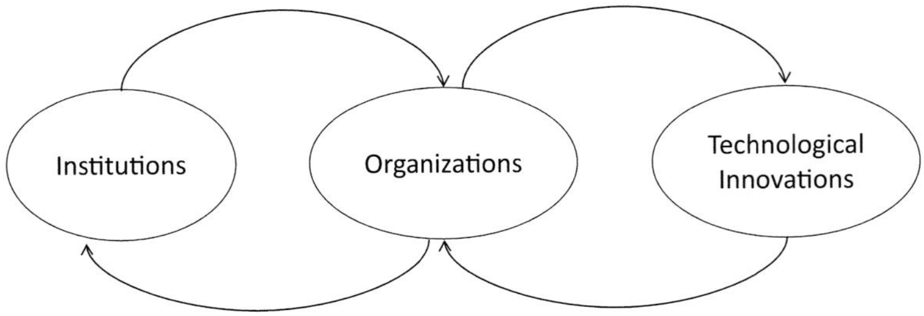

Fig. 10.1 Theoretical framework built upon the systems of innovation and co-evolution of technology and institutions literature (Authors' elaboration)

such as the multi-level perspective (MLP). In this perspective, regime transitions occur through an interplay of three key layers: the niche level, in which radical innovations are created; the socio-technical regime level, in which organizations interact under a set of rules; and the socio-technical landscape level, which is the macro-level and includes the environment that might impact the regime and the niches (Geels 2005b). One could draw parallels between the framework we propose in this paper and the MLP approach. The development of technological innovations in our framework might indeed coincide with the MLP niche level, while interactions between organizations might coincide with the MLP regime layer, and institutional changes can be seen as occurring in the MLP landscape layer. While our framework suggests that interactions between these three layers are not unidirectional and do not follow any pre-established hierarchies, the MLP perspective has been criticized for favoring bottom-up change models. Indeed, according to Berkhout and colleagues, the MLP approach tends to "*emphasize processes of regime change which begin within niches and work up, at the expense of those which directly address the various dimensions of the sociotechnical regime or those which operate 'downwards' from general features of the sociotechnical landscape*" (2004: 62). This is the first of three reasons why we chose not to use it for this study. The second reason relates to the depth of the studied phenomenon. The MLP approach looks at the transition from one regime to another through the replacement of specific technologies (Geels 2004), such as the replacement of horse-drawn carriages with the advent of the combustion engine (Geels 2005a). The case of smart card system replacement does not, in our opinion, have sufficient depth to be analyzed using MLP and does not (yet) relate to a regime change per se. The evolution and implementation of smart ticketing may well be part of a greater socio-technical transition, supported by ICTs, but we believe that adopting the MLP approach when analyzing only a tiny aspect of this larger transition would be a misuse of the framework. Finally, use of the MLP approach is reliant on the ability to look back several decades (Geels 2005b), which is not possible for the case presented here. In the following section, we present the historical evolution of the ticketing system in London, from 2002 to nowadays, before analyzing it using the framework proposed here. Figure 10.1 is a graphical representation of our theoretical framework.

10.3 A Tale of Two Smart Card Systems: The Oyster and Contactless Payment Cards

In 2002, London had 7.3 million inhabitants. At that time, private motorized transport accounted for 45% of the 22.9 million trips taken every day within London (TfL 2009a). In order to decrease this number, and encourage more urban dwellers to shift from private motorized transport to public transport and human-powered mobility modes, the mayor of London developed and implemented, jointly with Transport for London (TfL), the city's dedicated transportation authority, a set of specific policies. Chief among them, the development of the smart ticketing system became famous around the world.

In 2001, the first Mayor's Transport Strategy was released. The key proposals listed were to freeze bus fares, limit underground fares, introduce smart cards, and simplify ticketing across all transportation modes in London, including in National Rail (GLA 2001). At the end of the same year, TfL expanded the Mayor's Transport Strategy and presented its 2002/2003 business plan, in which its primary objectives included to reduce traffic congestion, increase public transport usage, improve network integration, and improve access to the transport system (TfL 2002a). At the end of 2002, TfL released its next business plan (2003/2004), which included a proposal for the development of a dedicated ticketing and boarding strategy that proposed the introduction of cashless operations in central London and an extension of the entire bus network by 2008 (TfL 2002b). Correspondingly, dedicated ticket barriers, machines, and validators were installed and TfL staff piloted the system for several months. Finally, in 2003, the city launched its new "Oyster" smart card system. The system supported both annual and monthly tickets and used the NXP/Philips' MIFARE Classic microchip. Using NXP's own proprietary 48-bit encryption technology, it was also able to store 1024 bytes of data (Muller 2016). The card was accepted on buses and in the underground, supporting free transfer between and within both modes.

In 2004, TfL launched the Oyster Pay-As-You-Go (PAYG) system (TfL 2009b) (initially named "pre-pay"). The idea behind this novel concept was to offer an alternative to seasonal passes. Instead, users would top-up money on their card, up to £90, from which transport fares would be deducted at the point of use. In 2004, TfL announced a new fare package that aimed to boost the use of Oyster cards. People using smart cards would not be impacted by the 2004 fare increase and could continue to pay the 2003 fare (TfL 2004). This was followed, in 2005, by the introduction of daily caps for PAYG Oyster cards (TfL 2005) and the extension of the Oyster network to National Rail services within Greater London. The extension was not completed until 2010, following the signing of an agreement between TfL and the Train Operating Companies in October 2009 (TfL 2011). The Oyster card was met with wide approval among users, and by 2006, was being used for more than 73% of the 9.2 million trips made daily on the city's public transport system (TfL 2007a).

In 2006, TfL launched the Future Ticketing Project (FTP), which had two main goals. The first was to gain a deeper understanding of the costs of fare collection incurred through the Oyster card system, and the second was to investigate other emerging payment technologies to assess whether it might be possible to reduce the costs of the Oyster system. It was estimated that in 2005/2006, revenue collection costs accounted for 14 pence out of every pound collected (TfL 2016a). This was followed by the delegation of the Mayor's Transport Strategy to TfL in 2009 (TfL 2009c). The document included a proposal (number 123) for the use of new technologies in order to reduce the cost and improve the efficiency of fare collection (Mayor of London 2010).

In 2009, a proposal for contactless payment cards (CPCs) was presented to TfL's board. The proposal outlined a five-step development plan that would call for the city to transition toward this new ticketing system. CPCs are credit cards that are equipped with a microprocessor and internal memory enabling communication between the cards and smart card readers through radio frequency technology. The chips embedded in the contactless payment cards are able to handle, store, and grant data access through interaction with the card readers, and they are also able to perform internal functions such as data encryption (Smart Card Alliance 2004). TfL was interested in CPCs both for their potential to greatly reduce the costs of fare collection, but also because the CPC systems could increase user convenience. Here, fares would be deducted directly from the users' bank accounts, so customers would not waste time topping up their Oyster cards.

The first two phases of the 2009 business plan were dedicated to introducing CPCs in the bus system (in 2012) and the underground (in 2012–2013). In 2013 weekly capping was to be introduced, and the Oyster card was to be decommissioned by 2015 (London Assembly 2011). However, due to malfunctions with the Oyster system, TfL decided to terminate the contract with the ticketing system operator, Transys, early, and in 2010 it enacted the break clause (TfL 2010), choosing instead to work with Cubic. Correspondingly, through the Future Ticketing Agreement (FTA), Cubic was charged with operating the smart card system, intellectual property rights, and systems asset ownership was given back to TfL. Part of Cubic's responsibility under this new contract was to design and supply new card readers that could support both the old model of Oyster transactions and the new CPC system. By 2012, TfL had issued more than 43 million Oyster cards; at this point, more than 80% of trips made on TfL's network used the Oyster card (TfL 2012a). The first operational phase of the FTP was launched December 2012 and consisted of enabling CPC payment on buses (TfL 2012b).

In 2014, after a competitive bidding process, TfL awarded Cubic the seven-year "Electra Contract," worth approximately £660 million, to manage and maintain the TfL ticketing system, which included both Oyster cards and CPCs. The contract was expected to bring a savings of about £11 million per year (TfL 2014). Since mid-2014, CPCs have been accepted on all TfL services and have been equipped with a weekly and daily cap (DfT 2014). Cash payment was terminated on buses in 2014, meaning that smart cards (Oyster and CPC) are now the only accepted means of payment for public transportation in London. In 2015/2016, CPCs accounted for

26% of all Pay-As-You-Go tube and rail journeys and for 23% of all bus trips (TfL 2016b). The Oyster card accounted for the remaining percentage.

By 2015, London's population had reached 8.6 million, overtaking the 1939 population record and cementing its position as Europe's most populous city. It seems unlikely that London's population will stagnate, as it is expected to grow to 10.2 million people by 2039 (GLA 2015). Out of the 26.6 million trips taken daily in 2014, 45% used public transport and 32% used private motorized transport (TfL 2015). In 2007, for the first time in the city's history, the number of trips taken using public transport was greater than the number of trips using private motorized vehicles; public transport has remained the dominant mode of transportation ever since.

Although smart ticketing systems cannot be considered as the only reason behind the increased public transport ridership, as other measures such as the congestion charge scheme and the development of new transport infrastructures that impacted private automobiles ridership were introduced, their impact on the public's shift from private motorized to public transport cannot be neglected. Based on their success, TfL and Cubic developed a license, worth up to £15 million, that enabled other cities to adopt London's contactless ticketing system technology. In October 2017, the Metropolitan Transit Authority in New York City awarded Cubic a contract to update its entire ticketing system and implement CPC as a new ticketing mean, aiming to phase out the traditional MetroCard by 2023 (Barron 2017).

In the next section, we use the theoretical framework presented previously in order to analyze this case and to gain an understanding of the interplay between the involved organizations, institutions, and innovations. That is, how did the institutions influence the behavior of the organizations that shaped the technological innovations? And how did these technological innovations influence the behavior of the actors who then influenced the evolution of the institutions?

10.4 Analysis

The analysis presented here is based on the theoretical framework introduced previously, through which we aim to understand the interplay between institutions, organizations, and technological development. Each of these interactions is represented by a capital letter between parentheses, as illustrated in Fig. 10.2.

The first informal institution that must be mentioned in the context of the Oyster card development is the tendency for the British public authorities to outsource service provision to private organizations, especially in the transportation sector. There are many examples of the private sector's long-standing involvement in transport-related activities. For instance, throughout the twentieth century, the London Underground experimented with various public and private ownership models. Having been under private ownership until 1933, it turned to public ownership until 1984 and partial privatization until the early 2000s, before finally

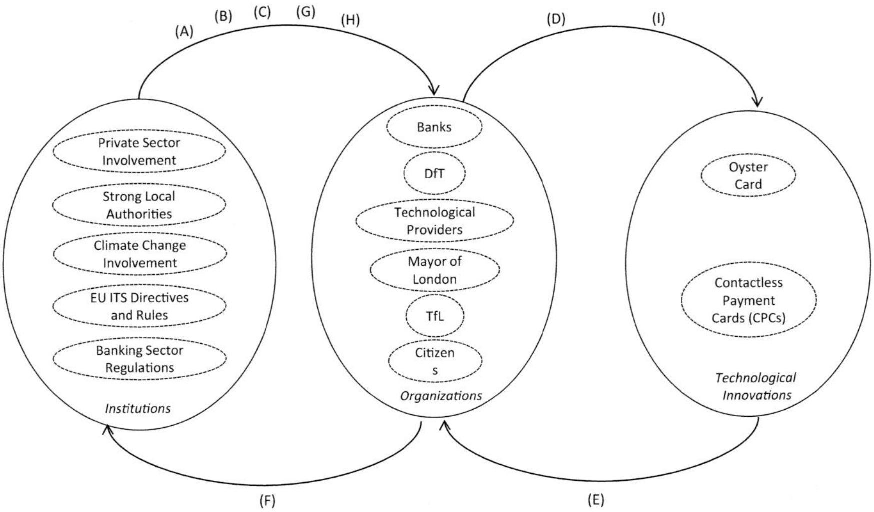

Fig. 10.2 Dynamic interactions between institutions, organizations and technological developments for the case of smart ticketing development in London

turning to public ownership. Similarly, the local bus system went from private ownership prior to 1933, to public ownership, before being finally re-privatized in 1986 (Preston 1999). The National Rail industry also followed a similar path, as it also went through successive waves of nationalization and privatization throughout the twentieth century (Finger 2014). We believe that the long-term involvement of private companies in the transportation sector has led to a certain institutional legacy that has itself impacted the involvement of the private sector in the development of the smart card schemes (A).

The fact that the UK is a decentralized country with strong local authorities, particularly in London, also likely had an impact on the transport system's evolution. Thus, we consider this significant level of autonomy to be another institution that influenced the development of smart ticketing schemes (B). As TfL, whose board is appointed by the mayor of London, is fully responsible for organizing transport services in London, it did not face any opposition from other actors, such as public authorities at the national level, when developing and implementing the smart ticketing systems.

The UK's climate change commitments also influenced the Oyster card development (C). For example, following the release of the 2000 Climate Change Program, which set national targets for the reduction of carbon emissions, the Department of the Environment, Transport and the Regions (DETR) (predecessor of the Department for Transport, DfT) published (building on the 1998 White Paper entitled "*A New Deal for Transport: Better for everyone*") its *Transport Act*, which aimed to promote solutions to fight congestion and pollution. Among other measures proposed in the Act, integrated ticketing systems are presented as a viable solution. In 2004, following the UK Emission Trading Scheme in 2002 and the subsequent publication of the Energy White Paper in 2003, the DfT released the Future of Transport Strategy, where smart cards were again presented as a means of attracting more public transportation users and fighting climate change. We can also argue that British involvement in climate change initiatives may have been influenced by the UK's EU membership, as all EU members are bound to emissions reduction targets fixed by the Commission. In any case, we believe that a combination of (A) and (C), and to a lesser extent (B), directly influenced interactions between public bodies and also pushed the public authorities to act quickly leading to the establishment of a private finance initiative (PFI) between TfL and the technological provider Transys (D) in 1998. This created the institutional climate that led to the development and implementation of the Oyster card.

The Oyster card has, in itself, also had an impact on organizations (E). In 2008, the European Commission released two action plans focused on intelligent transport systems (ITS): the 2008 Action Plan for the Deployment of Intelligent Transport Systems in Europe and the EU Action Plan on Urban Mobility, which included a section on ITS development (EC 2009). Along the same lines, a collaborative platform was created that aimed to facilitate a dialogue between EU member states on the implementation of ITS and to gather ideas about the future of ITS. It is through this platform that the UK public authorities presented their Oyster card solution and all its associated benefits associated to officials from other EU member states. In March 2010, the European Commission's directorate on Mobility (DG MOVE) organized a workshop for the development of the ITS platform (DG MOVE 2010), in which TfL participated. The EU ITS directive was released shortly afterward, in the summer of 2010. Thus, it seems fair to assume that these discussions (which occurred ex-post the release of the ITS 2008 Action Plan), where each of the states was invited to share their ITS experiences, played a role in the composition of the EU ITS directive. The UK authorities continued to play a role in defining the ITS vision and institutions at the EU level. For example TfL participated in the ITS Expert Group, where it published guidelines on technical or logistical aspects of ITS deployment in Europe and could therefore influence the way ITS had to spread out at EU level. TfL made itself irreplaceable for the European Commission when it came to smart ticketing. Thus, we can see how a technological innovation (the Oyster card) can influence the behavior of organizations (UK public authorities promoting smart ticketing at the EU level), while the organizations can simultaneously contribute to the definition and development of new institutions (F) with the EU ITS directive.

Likewise, the EU ITS directive also influenced the British public authorities in return as it demanded that each EU member state report on the current state of their ITS infrastructures as well as on its future plans; this could have put further pressure on TfL to move quickly toward CPCs. Thus, the UK's EU membership and resulting obligations (G), especially concerning transport digitalization, influenced the actors involved in the development and implementation of the CPCs as it likely pushed public authorities to move more quickly.

One formal institution that undoubtedly played a key role in determining the behavior of the actors involved in CPC development is the regulation of the UK banking sector (H). The rules and norms stipulated by the British banking sector left banks and card schemes sufficient freedom to act as real drivers in the development and implementation of CPCs. Under the CPC system, banks and card schemes would receive a transaction fee when their cards were used to pay for transport, which obviously meant they wanted the solution to be successful. Their participation was also supported by the fact that they were not tightly restricted by public authorities. For example, issuing banks did not face any opposition from authorities when they automatically provided their customers with CPCs each time their traditional credit cards expired; in this way, issuing banks started to equip their customers with CPC even before the new ticketing solution was implemented. For instance, in September 2007, Barclaycard (a subsidiary of Barclays) issued the OnePulse credit card, which was a contactless Visa card embedded with an Oyster card chip (TfL 2007b). According to the UK Card Association, by December 2016, 107.2 million CPCs had been issued in the UK. Although the OnePulse card was decommissioned in June 2014, it was a predecessor of the CPC and provides a clear example of the freedom afforded to the banking sector by British public authorities. We strongly believe that the combination of (A), (C), (E), (G), and (H) has directly influenced the definition of the FTA and the Electra contract in 2014, which led to the development and implementation of CPCs (I).

Last but not least, we can assume that both smart card systems have had an impact on the citizens, and more specifically on the interactions between the citizens and the London authorities. As both smart card systems were largely embraced by the public, one might assume, for example, that this could have had an impact on the elections of the mayor of London. However, as such a proposition is hard to verify and measure, we have decided to leave it as a supposition, and thus not to include it in Fig. 10.2. However, if verified, this would be synonymous of a new letter on the arrow from technological innovation to organizations.

Where it seems that the development of the Oyster was motivated primarily by climate change, cost savings, and the tendency of the public authorities to collaborate with the private sector, the CPCs also seem—despite also being motivated by cost savings for TfL—to have profited from flexible regulations in the banking sector, which allowed banks to take a pro-active role in the CPC roll-out. One could also summarize the above analysis by concluding that the Oyster development followed a more demand-pull approach than the CPCs, which seem to have followed more a technology-push approach.

10.5 Conclusion

In this chapter, we have proposed a theoretical framework to analyze the interactions between institutions, organizations, and technological developments in urban transportation systems, and applied the framework to the development and implementation of smart integrated ticketing schemes in London. From this analysis, it is evident that institutions influence the manner in which organizations interact with one another, which eventually leads to the birth of technological innovation. However, we have also shown the importance of considering the impact that technological innovations have on organizations, which can shape new institutional developments through their interactions. Therefore, it is essential that the interactions among organizations, institutions, and technology are considered dynamically.

How will the ticketing system in London evolve in the coming years? It seems fair to ask whether CPCs will eventually take over Oyster cards or whether both solutions will continue to coexist. Although TfL had originally planned to decommission the Oyster system by 2015, it seems that the London transport authority has been reconsidering its position. In 2009, TfL estimated that approximately 10% of its users, who probably correspond to lower socioeconomic statuses, were considered unbanked or underbanked, meaning that they did not have access to debit, credit, or prepaid cards (Brakewood 2010). Completely decommissioning the Oyster system and having the whole transport system rely only on the use of CPCs would mean that these people would be excluded from public transport, which fortunately did not happen. It also seemed that Londoners who were not unbanked or underbanked had grown attached to the Oyster card. Accordingly, to date, TfL has chosen not to decommission the Oyster system. On the contrary, in September 2017, TfL launched a dedicated app with a twofold purpose. Firstly, it gave Oyster users real-time information on their card's balance and their travel history. Secondly, it gave Oyster users the opportunity to top their cards up through the app via e-banking, which gave users the choice to pay for public transport directly with their credit cards or by charging their Oyster cards, which would then be used as a public transport ticket. Additionally, unlike the Oyster, CPCs are not capable of storing information of monthly, yearly, or discounted travel card-holders (Kamargianni et al. 2015). So even though CPCs have been shown to be cheaper for TfL than the Oyster card, it seems unlikely that the London transport authority will stop proposing both payment options.

Furthermore, since October 2016, London's black cabs have accepted CPCs (TfL 2016c), which can also be seen as an effective means of integrating public transport ticketing with taxis. Additionally, by creating a dedicated Web site to release most of its data to developers in 2009, TfL has also enabled third parties to create routing information apps that integrate different transportation means and enable users to access information about different transport options to better plan their trips. These two aspects (ticketing and information integration) are recognized as the building blocks for Mobility-as-a-Service (MaaS) (Kamargianni et al. 2016), which is presented as a solution "*combining services from public and private transport providers through a unified gateway that creates and manages the trip, which users can pay for with a single account*" (Civitas 2016: 2). According to Kamargianni and colleagues, a MaaS solution in London would benefit transport operators by creating a larger market, potentially benefit users by helping them to reduce their travel expenses and travel times, and help achieving a better modal distribution (Kamargianni et al. 2015, 2018). Thus, it might seem logical to keep MaaS in mind when considering potential evolutionary pathways for London's ticketing system.

Nevertheless, the viability of the smart card as a solution to the city's transportation issues still requires further exploration. It has been estimated that between 2002 and 2015 the proportion of London households without a car has remained stable at 41% (DfT 2016), and it has also been shown that for 2014–2015, the annual cost of traffic congestion in the Greater London Transportation road network equated to £5.5 billion (TfL 2016d). In June 2017 the mayor of London released a draft of the next Mayor's Transport Strategy for public consultation. According to the report, the population of London will be 10.5 million inhabitants in 2041, and if no further measures are taken to improve the conditions of the transport system, it is estimated that approximately 71% of travel on London underground and 65% of travel on National Rail will be crowded, which is defined as the number of standard class passengers surpassing the planned capacity for the service (Li and Hensher 2013).

The Oyster and CPCs are cited in the document as ways of making the London transportation system more pleasant to use, enabling customers to enjoy comfortable, confident, safe and secure, informed and stress-free travel. However, more coercive measures are proposed in order to reach targets of having the entire London transport system at zero emissions by 2050, or reducing the number of trips by private cars to 20% by 2041 (as compared to 36% in 2015). These new measures include the introduction of a road-charging scheme, which could lead to the removal of the congestion charge zones and different emission zones. ICT-based solutions, such as smart cards or, their more evolved versions such as MaaS, seem unlikely to induce societal changes, such as the transition from the automobility to the post-car paradigm, on their own. However, if embedded in the right set of policy measures and treated as an implementation tool rather than as an end in themselves, they might contribute to the construction of a more sustainable transportation system.

Acknowledgements The authors are indebted to Arnaud Besse-Cillier, Nicolló Ficarelli, Catarina Neves, and Theophile Verhnes for their assistance with background research for this paper, and to Rebecca Himsl for helping to edit the present paper.

References

Abrate G, Piacenza M, Vannoni D (2009) The impact of integrated tariff systems on public transport demand: evidence from Italy. Reg Sci Urban Econ 39(2):120–127

Audouin M, Finger M (2018) What can the South learn from the North regarding the implementation of IoT solutions in cities? The case of seoul-born smart transportation card implementation in Bogota. In: Hostettler S et al (eds) Technologies for development. Springer. https://doi.org/10.1007/978-3-319-91068-0_14

Audouin M, Razaghi M, Finger M (2015) How Seoul used the 'T-Money' smart transportation card to re-plan the public transportation system of the city; implications for governance of innovation in urban public transportation systems. In: Presented at the 8th TransIST symposium, Istanbul, Dec 2015

Barron J (2017) New York to replace MetroCard with modern way to pay transit fares. New York Times

Berkhout F, Smith A, Stirling A (2004) Socio-technological regimes and transition contexts. In: Elzen B, Geels FW, Green K (eds) System innovation and the transition to sustainability: theory, evidence and policy. Edward Elgar, Cheltenham, pp 48–75

Brakewood C (2010) Contactless prepaid and bankcards in transit fare collection systems. M.Sc. thesis, Massachusetts Institute of Technology

Brons M, Givoni M, Rietveld P (2009) Access to railway stations and its potential in increasing railuse. Transp Res Part A 43(2):136–149

Civitas (2016) Mobility-as-as-Service: a new transport model. CIVITAS Insight, 18

Crettenand N, Finger M (2013) The alignment between institutions and technology in network industries. Competition Regul Netw Ind 14:106–129

DfT (2014) Intelligent transport systems in the UK progress report as required by European Union Directive 2010/40/EU. http://ec.europa.eu/transport/sites/transport/files/themes/its/road/action_plan/doc/2014_uk_its_report_2014_en.pdf. Accessed 25 Jan 2017

DfT (2016) National travel survey: England 2015. https://www.gov.uk/government/uploads/system/uploads/attachment_data/file/551437/national-travel-survey-2015.pdf. Accessed 10 Mar 2017

DG MOVE (2010) Workshop intelligent transport systems for urban areas, summary of proceedings. https://ec.europa.eu/transport/sites/transport/files/themes/its/road/action_plan/doc/2010_03_18_summary_of_proceedings.pdf. Accessed 16 Oct 2017

Doloreux D (2002) What we should know about regional systems of innovation. Technol Soc 24(3):243–263

Edquist C (1997) Systems of innovation: technologies. Institutions and Organizations, Pinter/Cassell, London

Edquist C (2001) The systems of innovation approach and innovation policy: an account of the state of the art. In: Presented at the DRUID conference, Aalborg, June 2001

European Commission (2009) Action plan on urban mobility. http://eur-lex.europa.eu/legal-content/EN/TXT/PDF/?uri=CELEX:52009DC0490&from=EN. Accessed 25 Jan 2017

Fagerberg J, Verspagen B (2009) Innovation studies—the emerging structure of a new scientific field. Res Policy 38(2):218–233

Finger M (2014) Governance of competition and performance in European railways: an analysis of five cases. Utilities Policy 31:278–288

Finger M, Crettenand N, Laperrouza M, Künneke R (2010) Governing the dynamics of the network industries. Discussion paper series on the coherence between institutions and technologies in infrastructures, EPFL

Geels FW (2004) From sectoral systems of innovation to socio-technical systems: insights about dynamics and change from sociology and institutional theory. Res Policy 33(6):897–920

Geels FW (2005a) The dynamics of transitions in socio-technical systems: a multi-level analysis of the transition pathway from horse-drawn carriages to automobiles (1860–1930). Technol Anal Strateg Manag 17(4):445–476

Geels FW (2005b) Co-evolution of technology and society: the transition in water supply and personal hygiene in the Netherlands (1850–1930)—a case study in multi-level perspective. Technol Soc 27(3):363–397

GLA (2001) The mayor's transport strategy highlights. http://plangate.no/areal-transport/london/Transport%20for%20London%20-%20transport%20strategy%202001%20-%20highlights2.pdf. Accessed 27 Jan 2017

GLA (2015) Historical population 1939–2015. https://data.london.gov.uk/dataset/population-change-1939-2015. Accessed 16 Feb 2017

Goldman T, Gorham R (2006) Sustainable urban transport: four innovative directions. Technol Soc 28(1):261–273
Hirsch L, Jordan J, Hickey R, Cravo V (2000) Effects of fare incentives on New York City transit ridership. Transp Res Rec J Transp Res Board 1735:147–157
Huwer U (2004) Public transport and car-sharing—benefits and effects of combined services. Transp Policy 11(1):77–87
Janic M, Reggiani A (2001) Integrated transport systems in European Union: an overview of some recent developments. Transp Rev 2(4):469–497
Kamargianni M, Matyas M, Li W, Schäfer A (2015) Feasibility study for "Mobility as a Service" concept in London. UCL Energy Institute
Kamargianni M, Matyas M, Li W, Schäfer A (2016) A critical review of new mobility services for urban transport. Transp Res Proc 14(2016):3294–3303
Kamargianni M, Matyas M, Li W, Muscat J (2018) Londoners' attitudes towards car-ownership and Mobility-as-a-Service: impact assessment and opportunities that lie ahead. MaaSLab—UCL Energy Institute Report, Prepared for Transport for London
Li Z, Hensher DA (2013) Crowding in public transport: a review of objective and subjective measures. J Public Transp 16(2):6
London Assembly (2011) The future of ticketing. https://www.london.gov.uk/moderngov/documents/s7401/The%20Future%20of%20Ticketing%20-%20Appendix%201.pdf. Accessed 25 Jan 2017
Malerba F (2002) Sectoral systems of innovation and production. Res Policy 31(2):247–264
Markantonakis K, Tunstall M, Hancke G, Askoxylakis I, Mayes K (2009) Attacking smart card systems: theory and practice. Inf Secur Tech Rep 1(2):46–56
May AD, Roberts M (1995) The design of integrated transport strategies. Transp Policy 2(2): 97–105
May AD, Karlstrom A, Marler N, Matthews B, Minken H, Monzon A et al (2005) Developing sustainable urban land use and transport strategies. A decision makers' guidebook, 2nd edn. Institute for Transport Studies, Leeds
Mayor of London (2010) Mayor's transport strategy. https://www.london.gov.uk/sites/default/files/mts_executive_summary.pdf. Accessed 25 Jan 2017
Midgley P (2009) The role of smart bike-sharing systems in urban mobility. Journeys 2(1):23–31
Muller S (2016) How does an Oyster Card work? Alphr. http://www.alphr.com/technology/1002164/how-does-an-oyster-card-work-0. Accessed 13 Feb 2017
Nelson RR (1993) National innovation systems: a comparative analysis. Oxford University Press, New York
Nikolaou K, Toskas G, Mizaras V, Basbas S (2000) The environmental aspect of integrated smart card based services in Thessaloniki. Frenesius Environ Bull 9:341–346
Opurum CN (2009) Evaluation of the impact of automated fare collection system on rail rapid transit: the case of New York. Ph.D. thesis, University of Leeds
Potter S, Skinner MJ (2000) On transport integration: a contribution to better understanding. Futures 32(3):275–287
Preston J (1999) An overview of public transport in the United Kingdom and forecasts for the new millennium. Transport Studies Unit Oxford University; TSU Publications
Sheller M, Urry J (2000) The city and the car. Int J Urban Reg Res 24(4):737–757
Smart Card Alliance (2004) The what, who and why of contactless payments. http://www.smartcardalliance.org/resources/pdf/CP_What_Who_Why_Final.pdf. Accessed 13 Feb 2017
TfL (2002a) Agenda of TfL board meeting of the 19th of March 2002. http://content.tfl.gov.uk/agendamar2002.pdf. Accessed 27 Jan 2017
TfL (2002b) Agenda of TfL board meeting of the 19th of November 2002. http://content.tfl.gov.uk/agendanov2002.pdf. Accessed 27 Jan 2017
TfL (2004) New year new fares. https://tfl.gov.uk/info-for/media/press-releases/2004/january/new-year-new-fares. Accessed 27 Jan 2017

TfL (2005) Fares and tickets. http://infolondon.uw.hu/tfl-fares-2005.pdf. Accessed 27 Jan 2017
TfL (2007a) Annual report and statement of accounts 2006/2007. http://content.tfl.gov.uk/annual-report-and-statement-of-accounts-06-07.pdf. Accessed 10 Mar 2017
TfL (2007b) 3-in-1 card brings Oyster convenience to Barclaycard customers. https://tfl.gov.uk/info-for/media/press-releases/2007/september/3in1-card-brings-oyster-convenience-to-barclaycard-customers. Accessed 24 Jan 2017
TfL (2009a) Travel in London, Report 1. http://content.tfl.gov.uk/Travel-in-London-report-1.pdf. Accessed 10 Mar 2017
TfL (2009b) TfL's contactless ticketing: Oyster and beyond. http://www.bcs.org/upload/pdf/tfl-sep09.pdf. Accessed 27 Jan 2017
TfL (2009c) Draft mayor's transport strategy statement of intent. http://content.tfl.gov.uk/draft-mts-statement-of-intent.pdf. Accessed 25 Jan 2017
TfL (2010) TfL secures ownership and control of Oyster brand and ticketing system. https://tfl.gov.uk/info-for/media/press-releases/2010/april/tfl-secures-ownership-and-control-of-oyster-brand-and-ticketing-system. Accessed 27 Jan 2017
TfL (2011) Rail and underground panel. http://content.tfl.gov.uk/Item08-Oyster-NR.pdf. Accessed 25 Jan 2017
TfL (2012a) Join in the celebrations across the capital this summer with a limited edition summer Oyster card. https://tfl.gov.uk/info-for/media/press-releases/2012/june/join-in-the-celebrations-across-the-capital-this-summer-with-a-limited-edition-summer-oyster-card. Accessed 24 Jan 2017
TfL (2012b) Operational and financial performance report—second quarter, 2012/13. http://content.tfl.gov.uk/q2-2012-13-operational-and-financial-performance-report.pdf. Accessed 25 Jan 2017
TfL (2014) Revenue collection system and associated services procurement. http://content.tfl.gov.uk/fpc-20140717-part-1-item12-electra.pdf. Accessed 25 Jan 2017
TfL (2015) Travel in London, Report 8. http://content.tfl.gov.uk/travel-in-london-report-8.pdf. Accessed 10 Mar 2017
TfL (2016a) Transport for London: new mobility. http://mracongres.nl/wp-content/uploads/2016/03/Shashi-Verma-Transport-for-London-New-mobility.pdf. Accessed 22 Mar 2017
TfL (2016b) Annual Report 2015–2016. http://content.tfl.gov.uk/tfl-annual-report-2015-16.pdf. Accessed 25 Jan 2017
TfL (2016c) All London black cabs to take cards and contactless payments from Monday. https://tfl.gov.uk/info-for/media/press-releases/2016/october/all-london-black-cabs-to-take-cards-and-contactless-payments-from-mond. Accessed 17 Oct 2017
TfL (2016d) Traffic Note 4: Total vehicle delay for London 2014–15. http://content.tfl.gov.uk/total-vehicle-delay-for-london-2014-15.pdf. Accessed 2 Mar 2017
Urry J (2004) The 'system' of automobility. Theor Cult Soc 21(4–5):25–39
Van Nes R (2002) Design of multimodal transport networks: a hierarchical approach. Ph.D. thesis, TU Delft

Maxime Audouin holds a Bachelor of Science in environmental engineering, and a Master of Science in energy management, both from Ecole Polytechnique Federale de Lausanne (EPFL), Switzerland. Since 2015, he has been enrolled as a Ph.D. student in the Swiss Post Chair in Management of Network Industries (MIR) at EPFL, under the supervision of Prof. Matthias Finger. He research interest lies in the area of urban transportation systems, urban governance, transportation policy and regulation, as well as technology and innovation management. In his dissertation, he looks at the role that public authorities play in the development of ICT-supported integrated mobility schemes.

Matthias Finger received his Ph.D. degrees in political science and adult education from the University of Geneva, Switzerland. He is known for his expertise in matters of regulation and governance of network industries (postal services, telecommunications, electricity, railways, air transportation, urban public transportation, water). He has been an Assistant Professor at Syracuse University (New York), an Associate Professor at Columbia University (New York), and a Full Professor at the Swiss Federal Institute of Public Administration (Lausanne). Since 2002 he has held the Swiss Post Chair in Management of Network Industries (MIR) at Ecole Polytechnique Federale de Lausanne (EPFL), Lausanne, Switzerland, and since 2010 has directed the Florence School of Regulation's Transport Area at the European University Institute in Florence, Italy.

Chapter 11
The Governance of Demand-Responsive Transit Systems—A Multi-level Perspective

Fariya Sharmeen and Henk Meurs

Abstract In the new-generation smart mobility paradigm, Mobility-as-a-Service (MaaS) systems have shown the most potential to offer integrated mobility platforms. MaaS is essentially an amalgamation of services tailored to match individuals' diverse travel demands, many of which are demand-responsive transit (DRT) or DRT-like, addressing the much-needed last-mile problem of transit systems. Managing a DRT system could be a simple first step in understanding the governance of MaaS, as the two types of system share the common features of flexibility and individualistic, online platform-based service provision. With that view, we deliberate this exploration here, reflecting on the governance of a DRT case study. We borrow from the schools of transition management and industrial economics to elaborate our understanding. Using the framework of the Multi-level Perspective (MLP), we investigate the drivers and barriers for the diffusion of DRT through the (transit) regime. The main drivers of DRTs are found to be the willingness of local governments and transit operators to implement DRT, whereas the main barriers stem from infrastructure, technology, and market practices. Alignment among the regime elements remains largely partial, which is consistent with our conceptualization of MaaS governance challenges. This research adds to the literature by offering a comprehensive foundation exercise to reflect on the governance of innovative mobility services.

Keywords Demand-responsive transit · Mobility-as-a-Service · Multi-level perspective · Socio-technological transition

---

F. Sharmeen (✉) · H. Meurs
Department of Transport and Planning, Faculty of Civil Engineering and Geosciences, Delft University of Technology, Delft, The Netherlands
e-mail: f.sharmeen@fm.ru.nl

H. Meurs
e-mail: h.meurs@fm.ru.nl

F. Sharmeen
Department of Geography, Planning and Environment, Institute for Management Research, Radboud University, Nijmegen, The Netherlands

M. Finger and M. Audouin (eds.), *The Governance of Smart Transportation Systems*, The Urban Book Series, https://doi.org/10.1007/978-3-319-96526-0_11

11.1 Introduction

11.1.1 Motivation

Socio-demographic transitions, peak car trends, general regard for environmental sustainability, and the popularity of shared economy and flexible lifestyle have provided a massive impetus for the reconfiguration of urban transportation systems. Facilitated by the breakthrough in the information technology and wireless network sectors, a niche for demand-based public services has been created. Subsequently, transport services have also been reconfigured to suit the flexible lifestyles of individuals. Among this genre of new mobility services, Mobility-as-a-Service (MaaS) systems have shown the most potential, promising an integrated flexible mobility platform.

MaaS is defined as a '*user-centric, intelligent mobility distribution model in which all mobility services are aggregated by an operator and supplied to users through a single digital platform*' (Kamargianni and Matyas 2017). Such an integration of mobility services with one-stop shop offers access to a number of options, such as public transport (PT), car and bike sharing, as well as carpooling. Often, an online platform (usually smartphone apps) enables the planning, booking/reservation, and payment of the requested services. MaaS is regarded as the most promising integrated mobility system—a paradigm shift in daily transportation (Jittrapirom et al. 2017a). Considerable research efforts have also been undertaken that point at a number of advantages of such systems for the primary customers, including personalized offers as well as easy transactions, ease of payment, and journey planning (Karlsson et al. 2017).

Despite having significant social, economic, and environmental benefits for society as well as for individuals, it is not easy to implement MaaS. Firstly, potential users may be reluctant to change their choice patterns due to habit persistence. Secondly, suppliers of transport services may be disinclined to affiliate with these platforms due to a potential loss of revenues and a perceived risk of losing market share due to increased competition between the suppliers affiliated with the platform. In addition, the adjustments required to achieve interoperability may be costly for the firms involved. Finally, public agencies may have to reconsider mode-specific regulations, subsidies, and approaches in order to stimulate the development of integrated mobility services.

Providentially, the introduction of demand-responsive transit (DRT) systems in addition to conventional public transport postulates a viable test case in the implementation of MaaS, as they touch upon all of the above predicaments in a lighter scale. MaaS essentially is an amalgamation of services tailored to match the diverse travel demands of individuals in the context of contemporary shared economy, many of which are DRTs or DRT-likes, such as car-sharing or bike-sharing schemes. DRTs are perhaps the most important component to achieve successful MaaS integration as they provide a constructive solution to the first- and last-mile problems and complement fixed transit lines in low-demand locales and in off-peak hours. Similarly, given that MaaS is a demand-driven integrated service, it has a plausible association with the operation and governance of DRTs. Managing flexible personalized services of DRT could serve as a first learning step to the governance of MaaS, as they share the common features of flexibility and individualistic, online platform-based service provision (at least such is the case of the first Dutch DRT pilot, Breng flex).

Moreover, we argue that in order to understand issues concerning the transition toward MaaS, we must understand how current transport systems are stabilized through various lock-in mechanisms. Consumer lifestyles and preferences are adapted within the existing supply. On the supply side, this is concerned with investments, economies of scale and networks, infrastructure (bus stops, garages, etc.). In addition, a number of institutional aspects are relevant, including discourses, power relations, and the political networks. These lock-in mechanisms create path dependence that may be difficult to change drastically (Martin and Sunley 2006). A gradual introduction of these disruptive transit systems may allow the understanding and transition to penetrate through the existing transportation system.

MaaS enables multiple service elements with diversified parameters, stakeholders, and ownership constituents to come together and co-exist as facilitated by MaaS platforms. Accordingly, diligent management and governance of the provisions is a formidable undertaking. Since governance has been shifting to less provision and more management of services and mobility is transitioning to less ownership and more usership domain (Docherty et al. 2017), such knowledge would be crucial in terms of providing insights to facilitate complex mobility management system of MaaS. Much of the success in this transition from fixed to flexible transit system depends on the knowledge transfer. Therefore, it is imperative to understand and evaluate the governance of DRT as part and parcel of managing flexible service provision. These are the primary motivations of the present research.

11.1.2 What Is DRT?

As the name suggests, a DRT is a demand-based public transit system. Essentially, it means transport on demand. As opposed to having fixed transit lines operating on fixed routes on a fixed schedule, this service operates when and where users demand it. This is a lucrative option not only for users but also for the PT operators and government to tackle increasing expenses on operating low-demand lines.

In economic terms, DRT offers a horizontal product differentiation in the public transit domain. Industrial economics literature defines horizontal differentiation as offering substitute products (or services) at various combinations of the attribute's proportions. It responds to the taste heterogeneity of the consumers based on which the proportionate combination of the attributes in a product/service varies A range of substitute products can co-exist in the market, simply because the combinations of attribute proportions appeal to different target consumers. Vertical product differentiation, on the other hand, refers to a class difference among the products where all attributes of the product correspond to a higher or a lower range (Gabszwicz and Thisse 1986). It is important to understand the product differentiation as it has considerable effects on the operation and management of the service industry. In fixed PT and DRTs, the proportion of the attributes (travel time, travel cost, waiting time, comfort, connecting time to access and egress modes, flexibility) varies to cater the taste heterogeneity of consumers. Therefore, we categorize this as a case of horizontal product differentiation. It may lift itself to a vertical product differentiability if and when there is a case of up or downgrading of all the travel attributes.

The concept of DRT dates back at least 55 years to when the first feasibility study of demand actuated service was conducted (Bauer 1971). DRTs have been operational in Finland (Kutsuplus) and USA (Via, Bridj, UberPOOL). Understanding the governance and operation of DRT in the context of transition theories is a novel approach, to which the present research aims to contribute.

11.1.3 Research Objectives and Approach

This research investigates the potential role and impact of on-demand transit system on the governance of MaaS, as part of a transition towards implementing a comprehensive MaaS system. The primary research question is: What is the role of DRT in facilitating the governance of this transition toward a more integrated and flexible end-to-end mobility services within MaaS?

We elaborate this by means of analyzing a case study—Breng flex—which operates in the region of Arnhem–Nijmegen, medium-sized twin cities in the southeast of the Netherlands. The first step of MaaS in the case study involves the introduction of DRT services in addition to regular fixed scheduled PT services. Various other modes and mobility services will gradually be affiliated within a MaaS platform, allowing for integrated booking/reservation and payment. In the final stages, the planning options may be added to the services of the platform provider. This order of development is determined in part by the specific target groups of the pilot: commuters and students are usually aware of the transport systems they want to use and demand easy access and payment systems to use such services. To investigate the above-mentioned question, we evaluate the governance of this DRT to understand its role and contribution to an integrated MaaS system.

The multifacetedness of MaaS governance dictates an inter-disciplinary approach in the scientific framework as well. Thus, we borrow from the schools of transition management and industrial economics to elaborate our understanding. We employ the Multi-level Perspective (MLP) put forward by Geels (2002, 2012), who identified three levels of evolutionary reconfiguration: niches, regimes, and landscapes. Using the framework of socio-technological transition management theory of MLP, we evaluate the drivers, barriers, and alignment of the regime elements. Data was collected through primary (interviews) and secondary (documents, reports) sources. The analytical framework (elaborated in Sect. 4.1) was included in the interview structure, including a brief elaboration of the conceptual framework (Fig. 11.1).

This study adds to the literature by offering a comprehensive framework to reflect on the governance of innovative mobility services with an application to flexible transport. This framework is used to assess the introduction of DRT as a new mobility service from a governance perspective, reflecting on the meaning and role of it toward the governance of MaaS. While most governance literature on

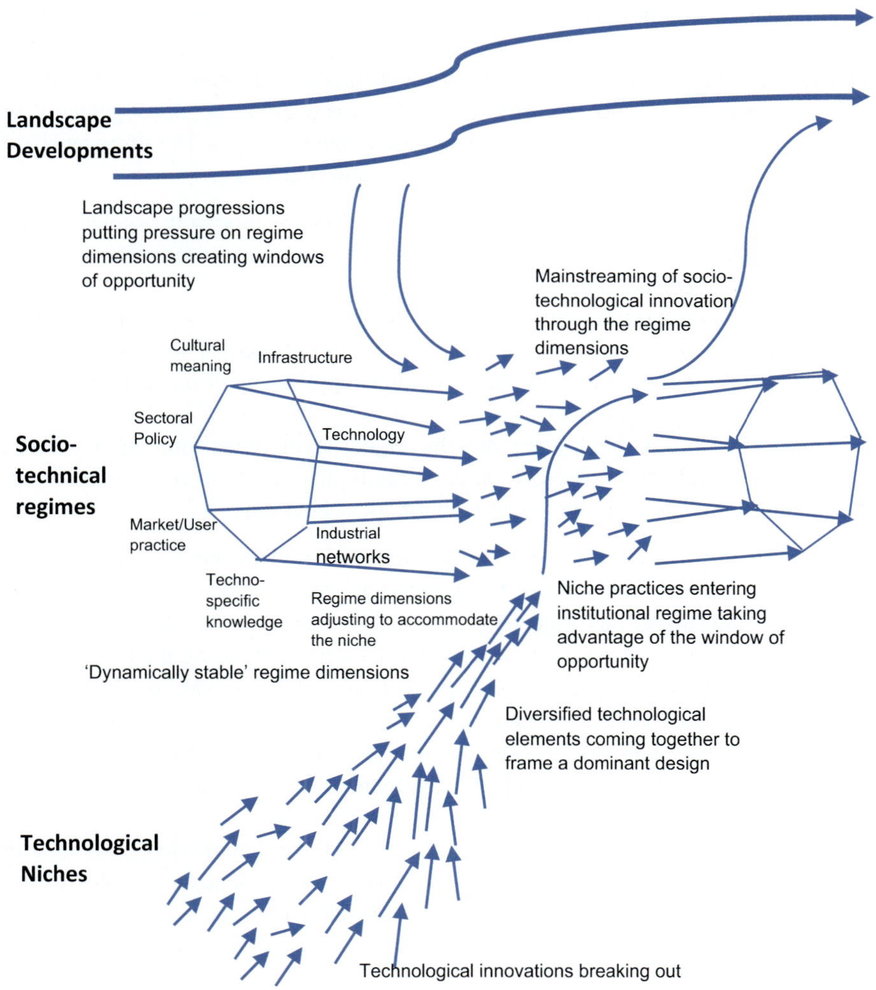

Fig. 11.1 Multi-level Perspective (MLP) of socio-technological transitions. (Based on Geels and Schot, 2007)

MaaS discusses different drivers and barriers with respect to the perceived ultimate integrated mobility propositions, we believe that a bottom-up understanding of the implementation of these services may be more fruitful in realizing a MaaS-system, taking all relevant actors into account.

11.2 Conceptual Framework

11.2.1 Multi-level Perspective (MLP)

As previously mentioned, the governance of DRT in this research will be examined using MLP introduced by Geels. This approach has emerged as a fruitful middle-range framework for analyzing socio-technical transitions to sustainability and has been used to inform on the governance or management of socio-technological transition (see an overview in Sect. 11.2.3). The three levels of the multi-level perspective are niches, regimes, and landscapes.

- Niches are technological incubators in which new innovative practice and breakthroughs are nurtured. They generate new practices and accommodate learning practices of radical innovations.
- Regimes accommodate such radical innovations (and generate incremental innovations). The process is gradual and can be explained by the trajectories of niche accumulation; that is, by the subsequent application of niche technologies at market or regime domains.
- Socio-technological landscapes are the higher-level representation of overall social, political, technological, and cultural change. Sometimes niches are triggered by changes in the socio-technological landscape.

The nested structure (Fig. 11.1) is important to understand the development trajectory of radical innovations from niche to mainstream. MLP provides a heuristic to understand this trajectory, where any change in one level affects the other. Success of new innovations is largely reliant on embedding them to existing regimes and socio-technological landscape. On one hand, innovations are nurtured in the technological niches; on the other hand, shift in landscapes create tensions demanding regime changes, making way for a 'window of opportunity' for the radical innovations to flourish. Thus, alignment and embeddedness are crucial elements of success of niche practices.

Although the ontological assumptions (Geels 2004) and methodological breadth have been criticized by scholars (Shove and Walker 2010; Genus and Coles 2008), we concur with Geels (2011) that the multi-dimensionality and hierarchical structure are particularly suited to understand the layered structure of socio-technological transitions.

If one considers the technological breakthroughs forming their niche somewhere at a micro level, positively fashioned, and in tune with the overarching sociopolitico-cultural landscape, regimes are the yielding ground through which the niches become mainstream. They can be considered as bridges between the homegrown socio-technological niches and the canopy of collective landscape. They frame, shape, and polish the niches to meet welfare goals, to uphold public values, and to become successful mainstream practice, with the ultimate goal of facilitating coherent societal transition.

This framework takes into account that transitions in mobility services are about interactions between technology, policy/power/politics, economics/business/markets, and culture/discourse/public opinion. In the present study, we employ the framework to understand the governance of DRTs. Governance here has been identified as a co-evolutionary cooperative management of the transition from the scheduled to demand-responsive transit systems. In the following sections, we will review the application of MLP in transportation research, followed by conceptualizing the case of DRTs within the MLP.

11.2.2 Application of MLP in Mobility

Everyday, mobility has been going through numerous episodes of transition over the recent years. Consequently, scholars have been resorting to transdisciplinary concepts of transition management to understand and explain the processes. Not surprisingly, several studies (Geels 2012; van Bree et al. 2010; Nykvist and Whitmarsh 2008) have utilized MLP to explain mobility transitions. Geels (2012) himself outlined how MLP can be used to explain low-carbon transition. He emphasized the usefulness of MLP to capture the co-evolutionary and multidimensional (involving technology, market, policy, culture) nature of the systematic transition that prevails in transportation. He identified the stability and the emerging cracks on the automobility regime introduced by climate change and environmental concerns. Coupled with intelligent transportation systems (ITS), alternative transition paths were identified to stem from potential niches of intermodal transport, bike/car-sharing, and demand management. Political will and user attitudes have been identified as the drivers of low-carbon transition and barriers to ITS.

In a similar context of low-carbon transition, van Bree et al. (2010) studied the transition of hydrogen-powered and electric vehicles, addressing emission and environmental concerns. They combined MLP with two scenario analyses on tightening emission standards and rising fuel prices. They also provide some insights into the dynamics of possible transition and discuss the relationship between industry (car manufacturers) and user (consumer) practices.

Similar to the scenario analysis approach (van Bree et al. 2010), Nykvist and Whitmarsh (2008) conceptualized a radical system innovation in transport through three niche routes: technological change, modal shift, and reduced travel demand. They showed the empirical evidence in the UK and Sweden on these three routes and explored the areas of convergence and contradiction. They identified landscape (environmental and economic) pressures and the response of regime dimensions as the necessary conditions of sustainable transition. Among the regime dimensions (Fig. 11.1), culture and infrastructure were found to be responding slowly to landscape developments.

Local spatial, cultural, and political factors have also been noted as keys to determining the transition pathways of similar transport initiatives (Marx et al. 2015). In a comparative study of Brazil and Germany, Marx et al. (2015) showed that different transition trajectory and niche growth patterns would be generated based on local factors, but also firm strategies. In other words, not only the speed but also the direction of technological niches is dictated by how the regime dimensions are swayed.

Moradi and Vagnoni (2018) identified traffic congestion, emission, and parking problems among the major cracks that have emerged from the landscape pressure and destabilized the automobility regime. They studied the driving and restraining role of urban mobility system dynamics and concluded that the drivers of the dominant (automobility) regime are the barriers to the public transit and non-motorized transit regimes and vice versa. Essentially, this implies that environmental concerns are the steering forces of the PT regime, as conceptualized in this study.

All of these studies have highlighted the multi-actor, multi-disciplinary, and co-evolutionary nature of transportation sector in one way or the other. What came across in all these studies is that mobility transitions can only be fostered through the mediation among all these actors and stakeholders. Based on the transition pathway, the actors can drive or constrict the socio-technological innovation. To summarize, it is imperative to investigate the response of regime dimensions to the landscape developments and an in-depth exploration of the drivers and barriers in order to understand the transition of socio-technological innovations.

11.2.3 Contextualizing DRT within MLP

Within the context of socio-technological transition and MLP, we contextualize DRTs where the niche is a new (flexible) mobility service, the regime is the regulatory and institutional domain, and the landscape is the changing perception and demand for flexibility in service provisions.

Niche: Breakthroughs in the IT and wireless network sectors have generated a niche for demand-based public services based on usage of interactive platforms, usually facilitated through smartphone apps (demand-based was already there using phones). This is in response to the dramatic increase in smartphone-based lifestyles where there are apps for just about everything nowadays. The impetus is the switch from fixed to flexible service to accommodate the flexible individualized work-life balance.

Regime: This is the regime of (public) transportation in shifting the regulatory environment within which the changes are taking place, for example, public service contracts or concessions using tendering procedures. There are several factors that could channel the regime shift. Firstly, who owns the idea, the data, and the service are important factors in regime development direction; for example, in Breng flex, is the concept owned by the transit authority that produced the idea or the local government that contracts (and perhaps also finances) the service? Finally, the legal framework could sway the regime development; for example, how the law makes a distinction between bus-services and taxi services could pave the way for DRTs to fit within this distinction, limiting the innovation.

Landscape: The present socio-political climate is characterized by socio-demographic transitions, flexible lifestyle, and changing perceptions toward car ownership ('peak car' trend), as well as a general awareness of sustainability. Regard for environmental sustainability (Moradi and Vagnoni 2018), coupled with antipathy of the use of fossil fuel and the popularity of circular economy, has fostered reconfiguration of transportation systems. At the same time, the rise of social media has led to a flexible lifestyle, mobile working arrangements, and shared economy. Recent socio-demographics trends have shown a general decline in car ownership (van Wee 2015; Oakil et al. 2016) and, consequently, more demand for usership and flexible demand-based individualized mobility services. All of these aspects paved the way for the appreciation of public transit and adoption of new forms of public transit that are more flexible and on-demand, yet also sustainable and speak to the preferences of car ownership and use.

Given that the technological niches are coming together to enable DRT to enter the public transit regime, we examine how the regime dimensions are accommodating this transition and responding to the landscape development-induced pressures. Within this context, the model goes on to examine on which dimensions the introduction DRT fits within current regimes and landscapes.

11.3 Case Study Description: Breng Flex

Breng flex, a DRT service operated by Dutch PT provider Connexxion, was conceived and proposed in response to the vision of new and smart mobility solutions of the province of Gelderland. The province set aside a budget for new mobility solutions and invited ideas from transport providers; among these, a DRT system was picked up that was later named as Breng flex. It is therefore financed by the regional government.

The first pilot was launched in December 2016; it was continued through subsequent pilot programs upon positive reception and evaluation (Alonso-González et al. 2018). Breng flex operates in the twin cities of Nijmegen and Arnhem in the Netherlands. It is essentially a bus-stop-to-bus-stop service (as opposed to door to door service) for most of the transit network, with some additional landmark stops noted as 'virtual bus stops' in the app platform; there are a total of 225 stops in the network. The service is territorial at this point of time and does not cross the city boundaries, which means that a consumer cannot use the service to travel between Arnhem and Nijmegen. It operates on a flat rate of EUR3.50 per person per trip regardless of the distance, travel time, or the socio-demographic profile of the user. Users can pay in advance via the smartphone app or on the vehicle using their debit cards or national travel cards (OV-chipcards). The operating hours are 06:30–24:00 on weekdays, whereas starting times are 8:00 and 9:00 on Saturdays and Sundays, respectively. Rides can be called and tracked in real time through the designated mobile app only.[1] Seats are guaranteed. The fleet comprises five minibuses (five passenger seats and two wheelchairs) and four electric cars (three passenger seats).

11.4 Analysis of Breng Flex Case

11.4.1 Analytical Framework

The analytical framework was built around the MLP concept. Geels (2012) identified seven dimensions of socio-technological regimes (Fig. 11.1)—infrastructure, market practice, sectoral policy, technology, industrial networks, techno-specific knowledge, and cultural meaning—which form the analytical framework for this research. Those dimensions (Table 11.1: column A) were then elaborated into elements (column B) specific to the case study. To structure the regime elements, we then introduced the notion of barriers and drivers of socio-technological innovation to the analytical framework (column C). For each of the elements, we carefully investigated the barrier to change and the drivers to success of the innovative mobility system. By doing so, we maintained a certain level of robustness to the analysis. Finally, the alignment among all the elements was evaluated based on the data analysis. A five-point Likert scale-based scoring system was used to report the findings, where 1 indicated not aligned at all and 5 meant completely aligned (column F). The scores represent how in or out of balance each of the elements is to the rest of them. Since Breng flex is not fully operational and running in the pilot phases, only where investment is one-way without any profit return, we have been restrained with the alignment scores. For example, even though certain elements seem to be devoid of any barriers, we did not assign those with a perfect alignment score of five.

Data was collected through primary (in-depth interviews) and secondary (documents, reports) sources. Four in-depth interviews were conducted among a representative of local regional government, an academic expert, a transit (DRT) operator, and a mobility consultant. The interviews conducted were structured based on the analytical framework, that is, broken down into regime dimensions and elements. The respondents were first familiarized with this framework in order for them to understand the context and to discuss the barriers and drivers accordingly. The interview recordings were then analyzed using the above-described analytical framework.

---

[1]The process is described in a short video in the Breng flex Web site (in Dutch) https://www.breng.nl/breng-flex/1411.

This method helps explain drivers and barriers with respect to next steps of MaaS, which, as a far more complex system, is likely to be disruptive. Understanding the complexities involved in balancing barriers and drivers for a single flexible service would provide a good illustration and foundation for more complex systems. A more elaborate discussion on the findings based on the regime elements is provided below. Here again, the same structure of regime dimensions is followed to report the findings cohesively.

11.4.2 Findings

The findings of the study are structured along the seven regime dimensions as follows.

Infrastructure: Breng flex builds on the established road public transit infrastructure of Connexxion, which is a strong driver of the system. It employs the existing designated bus stops as pick-up and drop-off locations. Where these are lacking, it uses landmarks and facility points such as retirement homes to pick up and drop off passengers. The fleet has been subcontracted out with the responsibility for managing and maintaining the vehicles. Because the fleet size is reasonably small, the existing parking facilities of Connexxion could sufficiently accommodate them as well.

A major incentive, as well as an obstacle, of the Breng flex initiative was network optimization, particularly to manage low-demand transit lines. While frequencies of some fixed transit lines have been reduced, others have been discontinued entirely. For example, the bus line connecting Nijmegen to the peripheral village of Oosterhout has been long under debate and eventually disappeared after Breng flex started, and the frequency of busses to the urban counterpart to Wijchen has been lowered. Therefore, network rationalization has been a crucial feature of Breng flex, which posed some challenges in terms of service delivery and marketing and has been tackled through market promotion (Box 1).

Box 1: Interview Excerpt on the Strategy to Rationalize Transit Network
Suspending bus line 3 to the village of Oosterhout and promoting Breng flex instead has been a challenge. To manage the transition and win user acceptance, the marketing team offered free rides to the residents of Oosterhout. Checking the legibility of residency was also made lenient. The strategy proved successful with a significant increase in popularity of Breng flex in the area.

(Marketing executive, Breng flex, February 2018)

Market Acceptance: Breng flex can be denoted as an improvement to existing PT system of the region with no spatial disparity (Alonso-González et al. 2018). Although only about 60% of the initial target of 600 trips per day has been

Table 11.1 Evaluation of socio-technical regime of DRT—case study Breng flex

Dimensions of socio-technical regime (A)	Elements of socio-technical regime of DRT (B)	Barriers and drivers (C)	Are the regime components aligned? (D)	Alignment score[a] (five-point scale; 1 not aligned, 5 = aligned) (F)
Infrastructure	Route network—articulated definition of service delivery	Existing + virtual bus stops; network opportunities; low-demand bus lines	Yes	4
	Logistics (vehicles, parking)	Subcontractor deal for vehicle supply and maintenance	Yes	3
	Cost structure (vs. quality and usability)	Too early to evaluate at pilot stage as heavily subsidized; business model in the making	Pending revision	2
Market acceptance	Market share	Less than expected (60% of target achieved); growth curve is steep	Partially	3
	Popularity/performance	Popular among car users, elderly, and rural counterparts	Yes	3
	Consumer board acceptance	Skeptical—complement PT not substitute	Partially	3
Sectoral policy	Central-local govt agreement: characteristic of system/regime: rules and regulation as WP2000	Governed by province; not much influence from capital or legal framework—pilot stage	Yes	4
	Regional transport policy and taxation/subsidies	Province finances pilot; negotiation on regular service budget section, definition, and tendering	Pending revision	2
	Public welfare (accessibility, equity) and service standards	Less attention to equity at pilot stage; costly for traveling with children; talks on inclusion of elderly, businessmen and low-demand lines	Pending revision	2
	Data privacy regulation	Contractual competition; data ownership; knowledge dissemination	Yes	3

(continued)

Table 11.1 (continued)

Dimensions of socio-technical regime (A)	Elements of socio-technical regime of DRT (B)	Barriers and drivers (C)	Are the regime components aligned? (D)	Alignment score[a] (five-point scale; 1 not aligned, 5 = aligned) (F)
Technology	IT/app development	Built on existing proof of concept (Abel)	Yes	4
	GPS track and trace	Bus lanes are missing in navigation maps	Partially	3
	Data storage and coverage	Struggling with dashboard—business intelligence	Partially	3
Industrial networks	Vehicle manufacturer	Vehicle supply and maintenance	Partially	3
	Existing private/shared taxi network	Limited resistance from Regio taxi—taxi market share is low	Yes	4
	Labor market/crew/collective labor agreement	Labor union conflict, debate, bad press	Partially	3
Techno-specific knowledge	IT literacy	Tackled with phone and desktop options	Partially	3
	Smart phone use/coverage/cost	Smart aids are being conceptualized	Partially	3
	R&D on demand-driven transit	Mostly internal and agile; some academic research	Partially	2
Cultural meaning	Perception	Perception of luxury service; emphasis on marketing and customer care	Yes	4
	Reliability and mutual trust	Vulnerable to IT	Partially	3
	Car dependency	Moderate, approachable	Yes	4

Note The table is a summary of findings of the data collected during this research on the pilot runs of Dutch DRT system Breng flex
[a]The scores are based on the analysis of the interviews conducted with the local government and DRT operator

achieved, the development curve has maintained a steep growth. Breng flex has also been successful in attracting car users to its user pool.

The service is popular among elderly people, who generally have a greater appreciation for the guaranteed seating arrangement. Moreover, users from the rural counterparts, where fixed PT lines are infrequent, also constitute a fair share of the user pool after some deliberation (Box 1).

User satisfaction was quite high, particularly in the survey conducted by the operators (score 8.2), as a result of which Breng flex also went on to win the Happy Travelers award of 2017.[2] However, an independent academic survey reports that while users are quite positive about the operating efficiency of Breng flex, they remain apathetic regarding the questions of equity and accessibility (Ali 2017). These user evaluations should be interpreted cautiously, though, as they were all conducted during the pilot runs, which were operated on a promotion and fixed trip fare of EUR3.50. Pending revision, this price will change quite substantially to sustain the service in the long run.

Sectoral Policy: Sectoral policy has been a driver of Breng flex. The increased attention given to reducing both carbon emission and the use of fossil fuel by the European Union and the national government has boasted the promotion of sustainable transit solutions. On a local level, the vision of Nijmegen being the European Green Capital of 2017 and a home for high-speed cycling infrastructures has further incentivized it (Kerr 2017; Sharmeen and Lagendijk 2017). As mentioned earlier, Breng flex was conceptualized as part of the mobility vision of the Province of Gelderland. The pilots are also financed by them.

The key component of governance of public services is to uphold those visions. In addition to maintaining carbon emission standards, governance would also maintain equity and justice (Martens 2016). Since individuals' capabilities and preferences are so heterogeneous, a combination of multiple ethical perspectives of distributive justice should be recognized (Pereira et al. 2017). Among those perspectives, the guided principle of distribution should be built on a more nuanced multidimensional framework of accessibility to meet the heterogeneous needs of individuals. DRTs could potentially serve the public transit domain to fill an important gap to match the need and capability diversity of people. Breng flex has shown some promise in verifying this notion. As mentioned, the service has become increasingly popular among the elderly population, which the transport operator has recognized as a budding target consumer group. On the flip side, it is inequitably pricey for families traveling with children and university students, as the pricing plan does not yet differentiate between socioeconomic groups. Under present Dutch PT subsidy regulations, children travel for free or at a discounted fare until the age of 12 and students receive a free university PT deal. These measures are crucial to maintain equity in service provision and to provide accessibility to facilities. The extent to which these measures are incorporated in DRT is a question of service definition and goal

[2]Detailed infographics of the survey can be accessed here (in Dutch) http://www.brengkenniscentrum.nl/blog/wat-vinden-reizigers-van-breng-flex/.

established by the provider with agreement to the local governance-related goals, and these have not yet been established in Breng flex.

On the other hand, some crucial elements of sectoral policy have not yet been addressed. The tendering regulation and data privacy guidelines for the fully operational phase are yet to be determined. Firstly, agreements are yet to be made about crucial procedural and administrative treads, such as whether open competitive tendering will be called for and whether DRTs should be budgeted through regular public transport fund. Secondly, data storage, privacy, and sharing are currently being controlled by the operator. There is less clarity about if and how the knowledge will be shared to comparable projects in other cities, and no guidelines have been set by the local and regional government regarding the privacy and storage of these data. These issues are primarily been regulated by general transit data guidelines. Those mandates need to be adapted and consequently updated for the platform-based service systems.

Technology: Breng flex defines routes by employing smart algorithms that were adopted from Abel.[3] Having a proof of concept of the technology provided a strong technological foundation for the system's efficiency. However, alignment with other data and analytical support systems remains a work in progress. For example, navigation maps miss out on bus lanes in some places, resulting in a mismatch between the app-displayed waiting and pick up time with the actual ones. Also, the business intelligence framework has not been developed to display a consistent dashboard. Such an imbalance among technological spheres is hampering service planning and reliability.

Industrial Networks: An important mobility component to meet the heterogeneous need of Dutch societies is the consumer-specific taxi provision serving specific groups like the elderly. Unexpectedly, such taxi associations did not act as a barrier for Breng flex, as the market share of taxis is low in the region. The network of vehicle manufacturers was also of less concern since the fleet was outsourced to another company.

A major barrier here was the Collective Labor Agreement (CLA) for the new service. The providers drew a CLA that is substantially different from that of the bus drivers of the fixed lines. Although the providers argued that the package is quite attractive, it faced considerable resistance from the socialistic labor union, which drew some negative publicity to it. The term 'flex' has a negative connotation to it when associated with labor agreement. Such conflicts are not uncommon when generally associated with the fear of reduction of employment and flexible labor contracts. Research suggests that, if managed properly, the interaction between industry reconfiguration and employment transition can lead to profit maximization when there is a larger share of cross-ownership of services (Fanti 2013),

[3]Abel was the first fully electrical fleet-operated shared taxi service based in Amsterdam, which was discontinued in 2017 (after operating for almost two-years) on account of market saturation of taxi services.

which is the case here since all the services belong to a parent company. However, given the existing shortage of bus drivers in the region, employment cuts do not seem to be likely. More importantly, Breng has managed to get jobs for people with low qualifications for the labor market as a starting position; some may become bus drivers in the future.

Techno-specific Knowledge: Techno-specific knowledge, literacy, and access are critical for the success of smart innovative solutions (Warnick 2001). This has been the key to the 'smart cities for smart citizens' debate. The smart mobility domain is no exception; therefore, the policy rhetoric should be supported by 'digital governance augmentation' (Wiig 2015). IT literacy and user affability remains a challenge for Breng flex as well, particularly reaching specific consumer groups. Breng flex's marketing team has been coming up with smart aids to address these challenges (Box 2). The research and development remained somewhat internal to Connexxion and agile with some academic research efforts (Haanstra et al. 2017; Alonso-González et al. 2018). Reports are mostly user-centric and remain positive for the pilot runs.

Box 2: Interview Excerpt on Reaching the Target Audience
The marketing team held presentations at nursing homes and elderly housing compounds on Breng flex, explaining user procedures to potential consumers. The interface was also extended beyond smartphones so that a ride can be called through the desktop computer at the reception of such facilities. Moreover, the potential of smart bracelets for these target consumers is also being explored.

(Marketing executive, Breng flex, February 2018)

Cultural Meaning: User perception, reliability, and image development are important components of any new innovative business development, and much attention has rightfully been paid to the product marketing, branding, and perception building. Breng flex is no exception. Proper marketing can respond to product differentiation; it can channel or eliminate false sense of popularity by creating the desired 'image' to the targeted consumers (Tremblay and Polasky 2002). In accordance with this notion, Breng flex has also strategized creative marketing.

Any new product or service will encounter some apprehension from consumers, mostly related to general uncertainty about efficiency and reliability. In Breng flex's case, there was a false perception that it was an expensive and luxury product. Coupled with the discontinuation of certain fixed lines, this created some negative discernment of Breng flex. As part of breaking this apprehension, a free welcome ride was offered to consumers.

Social networks were also used to bring in more consumers. Research suggests that social networks can stimulate novel choice options through the mechanisms of peer influence (Sharmeen 2015; Rasouli and Timmermans 2013). Social campaigns were launched where users can share a unique code to gain exposures through individual's social networks. Service accountability, loyalty to customers, and good customer service were ensured to maintain quality of level of service.

In an attempt to elaborate the governance of DRTs, an alignment score was estimated to each of the elements based on the analysis of the drivers and barriers in each of them. In order to effectively manage the transition of an innovation, a reasonable alignment among the seven regime dimensions is crucial. If one of them is far ahead or far behind, it would put the whole structure off balance. For example, if technology is not aligned with the sectoral policy, or if the cultural meaning is off balance with the techno-specific knowledge, governance of the socio-technological transition would hamper. It is evident from the findings that only a partial alignment among the regime dimensions has been achieved in Breng flex so far. This is not surprising since the DRT service is currently in the pilot phase and a fully operational business model is yet to be formulated. Nevertheless, it gives a good indication of the dimensions that needs attention for the governance of DRTs in the context of a Western country, namely techno-specific knowledge, sectoral policy, and infrastructure. Not surprisingly, these correspond to the three challenges of MaaS implementation noted in the introduction, that is, reluctance of users, disinclination among transit operators, and local government challenges.

11.5 Conclusion

In this research, we investigated the potential role and impact of on-demand transit system on the governance of MaaS, as part of a transition towards implementing a comprehensive MaaS system. From the school of transition theories, multi-level perspective (Geels 2002, 2012; Geels and Schot 2007) was used as the theoretical framework to understand regime resistance. The case of demand-driven transit (DRT) was explored through a case study in the Netherlands to understand and evaluate barriers, drivers, and alignment of the elements of the socio-technical regime. Based on the theoretical framework of MLP, seven regime dimensions were explored and analyzed using data collected through in-depth interviews and a literature review. The regime dimensions were further broken down into case-specific (DRT) elements, and the concepts of barriers and drivers were introduced to understand potential, uncertainty, and, most importantly, the alignment among the dimensions. It is imperative to understand how these dimensions are coherent in order to reflect upon the governance and drawing out possible strategic responses.

The findings of our case study suggest that the regime elements are only partially aligned with each other. In particular, responses of infrastructure, sectoral policy, and techno-specific knowledge remain marginal. Recall that the case study (Breng flex) is at its pilot stage now and the fully operational business model is yet to be finalized and agreed upon. Quite substantial learning elements have been achieved through the pilot runs that would add value to the dynamic adaptation of public transit systems.

Findings also suggest that the main drivers of DRTs are the demand for flexible and efficient transit options, the willingness of the local government and transit operators, the finances to back up the initial investment for operation, marketing, and research to comprehend demand and supply repertoires. On the other hand, the barriers stem from the readiness of IT, pricing strategies, market share, techno-specific knowledge of users, and labor market resistance.

Although there is limited scope for DRTs to argue a vertical product differentiation to road transit systems, it can certainly present a compelling case of horizontal product differentiation. Using proper market incentives, it can add the dimensions of personalization and flexibility to the road public transit systems. However, careful strategizing would be crucial here as it is a case of horizontal product differentiation; intra-industry trade may hamper economies of scale and labor participation (Aturupane et al. 1999). Achieving a stable market equilibrium with horizontally differentiated products is less frequent as it is more prone to preferential choice of consumers (Gabszwicz and Thisse 1986).

The upcoming business model for the fully operational phase of Breng flex should be carefully drawn taking all these factors into consideration including learning from the example of the discontinued shared taxi service Abel operated with a complete electric fleet in Amsterdam (van der Veer 2017), which was suspended in 2017 on account of cost inefficiency. It would be crucial to understand how to balance operating capacity and service delivery. A dynamic pricing strategy can be applied to sustain service delivery with creative incentivization to manage pick up and waiting time (Amirgholy and Gonzales 2016). Moreover, DRT providers can deploy pricing strategies based on the sustainability and environmental awareness concerns. A spatial duopoly model application demonstrates how general awareness for the environment could affect the pricing strategies and market shares of competing products (Conrad 2005).

DRTs could be an important part to enhance public transit service and ensure seamless connectivity by particularly serving low-demand lines and connecting low-density urban counterparts. It is, however, debatable how reliable these services would be if offered as a substitute to regular public transit on account of both efficiency and accessibility these may offer. The prospects in general are quite promising given proper spatiotemporal service coverage in terms of not only filling but as well as complementing public transit system to enhance accessibility.

The perceived role of DRTs could be a crucial first step to the integrated Mobility-as-a-Service system particularly with respect to acquiring knowledge and experience of operating and managing flexible transit systems. They can contribute by addressing people's heterogeneous flexible demand profile particularly to serve rural and off-peak counterparts of daily urban systems. Breng flex has showcased a viable example of DRT generating several spin-offs within the country as well. One rather key aspect, however, remained disregarded so far, which is knowledge sharing and dissemination. Development of a knowledge network would be a valuable platform not only as a source of information but also to document best practice examples. National legal frameworks could play a role in enabling that knowledge base.

DRTs present a foundation exercise to understand flexible transit management of MaaS. Having said that, we would reiterate that in MaaS, DRTs are one of many components. The success of MaaS depends on mutual coherence and alignment among all those components. MaaS would be much more extensive having a range of services supplied by public–private operators meeting the demands of diversified heterogeneous consumers. The complexity is quite high, as was also evident from operational MaaS platform evaluations (Karlsson et al. 2016). For example, in the evaluation of Swedish MaaS system UbiGo, (Karlsson et al. 2016) identified regulations and institutional mandate as one of the major barriers. This was not found to be a barrier for Breng flex. Therefore, the findings of this research can be indicative only and cannot be generalized for MaaS. A DRT system offers a practical example of managing a flexible demand-based transit system. The knowledge base acquired from such a case is quite valuable in terms of achieving an efficient MaaS system. To elaborate further, a DRT would require a single business model, whereas a MaaS would thrive on a business ecosystem fed by many such models. There is little likelihood of cross-ownership of services. A recent study on the potential MaaS system for the Nijmegen–Arnhem region has illustrated the complexities and uncertainties of using a dynamic adaptive policy-making approach (Jittrapirom et al. 2017b). Therefore, the business models need to be carefully thought through for the development of the business ecosystem (Ebrahimi et al. 2018; Kamargianni and Matyas 2017). Likewise, the interrelation among the stakeholders—that is, service providers, users, and the data/platform owners—should be carefully scrutinized for the success of MaaS (Meurs and Timmermans 2017).

This research offers a detailed evaluation of DRTs under the robust framework of multi-level perspective. It explores the perceived role and prospects DRTs in transitioning toward a flexible and integrated mobility ecosystem of MaaS, analyzing the barriers, drivers, and alignment of the regime components from a governance point of view. It contributes a mid-level perspective, adding to the user-centric evaluation research of DRTs. The study is limited to one case study at its pilot phase, with limited stakeholder involvement. Other case studies of fully operational DRTs in comparable contexts would provide further insights. MLP can also be used with a similar analytical construct to evaluate the governance of MaaS. This remains on the future research agenda as a plausible next step of this deliberation.

References

Ali A (2017) BrengFlex: how efficient is it from a user's perspective? Radboud University

Alonso-González MG, Liu T, Cats O, van Oort N, Hoogendoorn S (2018) The potential of demand responsive transport as a complement to public transport: an assessment framework and an empirical evaluation. In: 97th annual meeting of transportation research board. Washington D.C

Amirgholy M, Gonzales EJ (2016) Demand responsive transit systems with time-dependent demand: user equilibrium, system optimum, and management strategy. Transp Res Part B: Methodol 92:234–252

Aturupane C, Djankov S, Hoekman B (1999) Horizontal and vertical intra-industry trade between Eastern Europe and the European Union. Weltwirtschaftliches Archiv 135(1):62–81

Bauer HJ (1971) Case study of a demand-responsive transportation system. Highway Research Board, pp 171–124

Conrad K (2005) Price competition and product differentiation when consumers care for the environment. Environ Resour Econ 31(1):1–19

Docherty I, Marsden G, Anable J (2017) The governance of smart mobility. Transp Res Part A: Policy Pract

Ebrahimi S, Sharmeen F, Meurs H (2018) Innovative business architectures (BAs) for Mobility as a Service (MaaS)-exploration, assessment, and categorization using operational MaaS Cases. In: 97th annual meeting of transportation research board. Washington D.C

Fanti L (2013) Cross-ownership and unions in a Cournot duopoly: when profits reduce with horizontal product differentiation. Jpn World Econ 27:34–40

Gabszwicz JJ, Thisse J-F (1986) On the nature of competition with differentiated products. Econ J 160–172

Geels FW (2002) Technological transitions as evolutionary reconfiguration processes: a multi-level perspective and a case-study. Res Policy 31(8–9):1257–1274

Geels FW (2004) From sectoral systems of innovation to socio-technical systems: Insights about dynamics and change from sociology and institutional theory. Res Policy 33(6–7):897–920

Geels FW (2011) The multi-level perspective on sustainability transitions: responses to seven criticisms. Environ Innov Societal Transitions 1(1):24–40

Geels FW (2012) A socio-technical analysis of low-carbon transitions: introducing the multi-level perspective into transport studies. J Transp Geogr 24:471–482

Geels FW, Schot J (2007) Typology of sociotechnical transition pathways. Res Policy 36(3): 399–417

Genus A, Coles A-M (2008) Rethinking the multi-level perspective of technological transitions. Res Policy 37(9):1436–1445

Haanstra A-M, van der Pool E, van Weert A (2017) Eerste monitoring- & Evaluatierapportage Breng flex. HAN, Nijmegen

Jittrapirom P, Caiati V, Feneri A-M, Ebrahimigharehbaghi S, González MJA, Narayan J (2017a) Mobility as a Service: a critical review of definitions, assessments of schemes, and key challenges. Urban Plann 2(2):13

Jittrapirom P, Marchau VA, Meurs H (2017b) Dynamic adaptive policymaking for implementing Mobility as a Service (MAAS). Paper presented at the european transport conference, Barcelona

Kamargianni M, Matyas M (2017) The business ecosystem of Mobility-as-a-Service. In: 67th annual meeting of transportation research board. Transportation Research Board, Washington DC

Karlsson M, Sochor J, Strömberg H (2016) Developing the 'Service'in Mobility as a Service: experiences from a field trial of an innovative travel brokerage. Transp Res Proc 14:3265–3273

Karlsson M, Sochor J, Aapaoja A, Eckhardt J (2017) Mobility-as-a-service: development of a tentative impact assessment framework. In: International conference on Mobility as a Service (ICOMaaS), 28–29 Nov. Tampere

Kerr L (2017) A tale of two green cities. Exploring the role of visions in the development of green infrastructure in two European Green Capital Cities. Radboud University

Martens K (2016) Transport justice: designing fair transportation systems. Routledge

Martin R, Sunley P (2006) Path dependence and regional economic evolution. J Econ Geogr 6(4):395–437

Marx R, de Mello AM, Zilbovicius M, de Lara FF (2015) Spatial contexts and firm strategies: applying the multilevel perspective to sustainable urban mobility transitions in Brazil. J Clean Prod 108:1092–1104

Meurs H, Timmermans H (2017) Mobility as a service as a multi-sided market: challenges for modeling. In: 96th annual meeting of transportation research board. Washington DC
Moradi A, Vagnoni E (2018) A multi-level perspective analysis of urban mobility system dynamics: what are the future transition pathways? Technol Forecast Soc Chang 126:231–243
Nykvist B, Whitmarsh L (2008) A multi-level analysis of sustainable mobility transitions: niche development in the UK and Sweden. Technol Forecast Soc Chang 75(9):1373–1387. https://doi.org/10.1016/j.techfore.2008.05.006
Oakil ATM, Manting D, Nijland H (2016) Determinants of car ownership among young households in the Netherlands: the role of urbanisation and demographic and economic characteristics. J Transp Geogr 51:229–235
Pereira RH, Schwanen T, Banister D (2017) Distributive justice and equity in transportation. Transp Rev 37(2):170–191
Rasouli S, Timmermans H (2013) Influence of social networks on latent choice of electric cars: a mixed logit specification using experimental design data. Netw Spat Econ 1–32
Sharmeen F (2015) Dynamics of social networks and activity travel behaviour. Eindhoven University of Technology
Sharmeen F, Lagendijk A (2017) Cycling practices through the lens of innovation biographies. Paper presented at the Asia Pacific Cycle Congress, Christchurch
Shove E, Walker G (2010) Governing transitions in the sustainability of everyday life. Res Policy 39(4):471–476
Tremblay VJ, Polasky S (2002) Advertising with subjective horizontal and vertical product differentiation. Rev Ind Organ 20(3):253–265
van Bree B, Verbong GPJ, Kramer GJ (2010) A multi-level perspective on the introduction of hydrogen and battery-electric vehicles. Technol Forecast Soc Chang 77(4):529–540. https://doi.org/10.1016/j.techfore.2009.12.005
van der Veer D (2017) Public transport and mobility on-demand research on the efficiency of on-demand taxi service Abel. Radboud University
van Wee B (2015) Peak car: the first signs of a shift towards ICT-based activities replacing travel? A discussion paper. Transp Policy 42:1–3
Warnick B (2001) Critical literacy in a digital era: technology, rhetoric, and the public interest. Routledge
Wiig A (2015) IBM's smart city as techno-utopian policy mobility. City 19(2–3):258–273

Fariya Sharmeen is a researcher in transportation planning specialized in mobility, networks, and urban regeneration. Her research interests include travel behavior, active travel, ICT and social interactions, social networks, community development, and city regeneration. She is proficient in a range of quantitative, qualitative, and spatial analytic techniques and has extensive experience in survey design and data collection. She has worked in diversified academic and research institutes across Asia and Europe. Her research has been published in leading transportation journals. She has also contributed to several books on travel behavior and has received awards for her research contributions.

Henk Meurs is Professor of transport and spatial planning at Radboud University in the Netherlands. He conducts research into mobility and spatial development and mobility. He has experience in policy development and evaluation in areas such as road infrastructure, public transport, and cycling, on a wide range of spatial levels. He is affiliated with the SURF-program SCRIPTS. He is also director of the MuConsult consultancy firm in Amersfoort.

Chapter 12
The Role of Public Authorities in the Development of Mobility-as-a-Service

Yanying Li

Abstract Mobility-as-a-Service (MaaS) is emerging as a key trend for providing more sustainable urban transportation. Decades of car-centric developments have led to congestion and air pollution. Therefore, it is vital to shift to public transportation and other greener modes. Efforts to enable this shift are underway, including multimodal transportation information systems. More recent trends include ride-sharing platforms and travel-broker services. MaaS holds the promise of unifying these services and providing a single platform for information, booking, and payment for all modes of transportation, potentially as a subscription model. However, if the MaaS concept is not checked by regulators, it could also have negative effects, such as increased inequality or a shift away from conventional public transportation to (single occupancy) e-hailing. Policy-makers must respond to, and guide implementation toward, maximum public good. Four distinctive case studies were analyzed to give recommendations for the role of public authorities. Recommendations include developing a balanced governance model with clear key performance indicators (KPIs), careful consideration of evolving urban mobility and corresponding legislation and funding, and the need for a robust evaluation methodology covering environmental and economic aspects.

Keywords Mobility-as-a-Service (MaaS) · Urban mobility · Public authorities
Business models

12.1 Problem Formulation: Urban Mobility and Modal Shift

A modal shift from private cars to public transportation can bring significant benefits to a city, such as reduced congestion, more reliable travel times, and reduced air pollution. Reduced use of private cars also has significant impacts on

Y. Li (✉)
European Climate Foundation, Brussels, Belgium
e-mail: yl3.soton@gmail.com

M. Finger and M. Audouin (eds.), *The Governance of Smart Transportation Systems*,
The Urban Book Series, https://doi.org/10.1007/978-3-319-96526-0_12

land use, such as saved space for building new roads or parking infrastructure (Banister 2008). It is a policy priority in the EU to make public transportation more convenient and comfortable to facilitate a modal shift (EC 2009). It has been long recognized that quality of public transportation service can influence car ownership and usage of public transportation (Goodwin 1993; Davison and Knowles 2005). High-quality public transportation service is particularly important today, as owning a car is no longer necessary, due to changes in lifestyles of younger generations and urbanization (Wharton 2017). Although such social trends are bringing new challenges to the transportation sector, they are also providing many opportunities to make cleaner and more sustainable urban mobility. As a result, new transportation modes, business models, and players in the sectors are emerging (ITF-OECD 2016).

A journey in a city commonly involves several transportation modes by various operators, who all have their own travel information services and individual tickets. Facilitating multimodal journeys has been a key transportation policy at city, regional, and national levels (DfT UK 2013; EC 2017a). Since the 1990s, public transportation operators and authorities have made considerable efforts to provide multimodal information and integrated ticketing to allow travelers to use a single information source and a single unified ticket to access transportation services from various operators in the same city or region (Dotter 2016). In almost any city, season tickets covering access to bus, metro, and local railway services operated by different companies are available. In addition, many individual countries (House of Parliament UK 2014), as well as the EU, have heavily invested in developing nation- and EU-wide traveler information systems to encourage and assist public transportation operators to open their data and application programming interfaces (APIs), under the EU's ITS Directive 2010/40/EU (EC 2017b).

The open-data policy enables a third party to develop travel information services, which traditionally is a service provided by transportation operators. This policy has successfully catalyzed travel information service markets, resulting in the availability of various information services in a given city. Taking Helsinki as an example, the market of apps for journey planners in Helsinki has been booming since public transportation data has been available to third-party developers (Li and Kristensen 2014). Such efforts to open data have not only improved passengers' experiences with public transportation, but have also stimulated innovations in the transportation sector.

Over the last decade, disruptive innovations have been transforming daily life in many aspects. A key example of this in the transportation sector is the wide availability of various modes of shared mobility, either bicycle/carsharing (shared vehicle) or ride-sharing (shared access). This has transformed urban mobility by providing more flexibility and convenience to users, ultimately discouraging car ownership (Iacobucci et al. 2017). Although shared mobility is not a new concept (shared bicycles have been available since the 1960s), the real boom of shared mobility only recently started with the increasing use of smart phones, ubiquitous Internet access, and social networks, since public transportation may not always meet the demands of travelers. Shared mobility has emerged in the market to meet such demands.

Many cities have integrated shared bicycle and public transportation services by using a public transportation card to access bicycles (Shaheen and Guzman 2011). Bicycle-sharing is often provided or subsidized by municipalities, as its benefits for a city's transportation are well recognized (Midgley 2011). However, other shared mobility services are often provided by private companies, such as e-hailing (e.g., Uber and Lyft) and carsharing providers (e.g., Car2Go and DriveNow). These are rarely included into a city's integrated public transportation information and ticketing systems.

In some US towns and cities, partnerships between public transportation companies and transportation network companies (TNCs) have formed, with varying setups and objectives. Many of these partnerships are aimed at solving the last-mile mobility issue by providing easy access to bus stops in a certain area. For example, the Pinellas Suncoast Transit Authority (PSTA), which provides public transportation for Pinellas County, Florida, launched a program called Direct Connect, with Uber and United Taxi, which provides Uber services between places of residence and bus stops (Uber 2016). Since 2017, all trips taken to or from the bus stops of PSTA will be $5 off the regular UberX price. In that way, shared mobility services have become an integrated part of the local transportation system. A study from the American Public Transportation Association (APTA 2016) concluded that e-hailing can encourage more people to use public transportation, enabling a modal shift to public transportation from private cars. With appropriate policy frameworks, e-hailing can substitute or complement public transportation rather than a competitor (Hall et al. 2017).

In 2011, the Association of Public Transportation operators (UITP) forecasted that combining various transportation modes (such as car-, taxi- and bicycle-sharing, car-pooling, and demand-responsive transportation) can complement traditional fixed lines/routes of public transportation (UITP 2011). New transportation services available in the personal mobility market have presented great opportunities to a new type of player, referred to as a travel broker (Sochor et al. 2014). A travel broker does not provide transportation services, but integrates various transportation services and sells them to end users. A MaaS provider acts as s a travel broker who provides integrated services to end users, providing the maximum levels of the convenience of the travel experience, which can give users similar, or even better experiences than using private cars.

Over the last few decades, the emergence of MaaS has been enabled by progress in both transportation policies and technologies, such as multimodal journey planners, shared mobility concepts, open-data policies, travel information apps, and e-ticketing. Despite these potential benefits, concerns have been raised over MaaS' increasing inequality, in which premium service levels are offered to those who pay more. This can be due to de-incentivizing sustainable mobility (POLIS 2017). The success of new services in some markets, including apps for private-hire vehicles and e-hailing, has clearly demonstrated the potential to disrupt existing urban mobility services and could encourage a shift toward car use, away from more sustainable modes (EPOMM 2017).

Since 2015, there have been various debates into what constitutes MaaS, involving providers, public transportation operators, public authorities, and academia (Kamargianni et al. 2015; Goodall et al. 2017). The current hype over MaaS, jointly driven by technology and business priorities, is beginning to affect policy thinking and policy-making at various levels. It is important that public authorities (at city, regional, and national levels) who play a key role in regulating, providing, and financing transportation services contribute to this debate. Society's ability to capture the benefits and minimize negative impacts of MaaS depends on the presence of effective policies and regulatory frameworks. A key question for public authorities is what kind of roles they should play to maximize the benefits and minimize the negative impacts from MaaS?

To answer the question, a review of the literature pertaining to the evolution from the first travel information systems and brokers, to the current MaaS concept, will give valuable background and first key insights. Four distinctive case studies were analyzed to understand the roles public authorities have played and the associated effects. Key enablers of deploying MaaS were also identified through the case studies, providing a framework for developing recommendations for the role of public authorities.

12.2 Literature Discussion: From Travel Brokers to MaaS

The concept of MaaS emerged in 2015. Since then, many projects and organizations have embraced the concept, finally making it a reality. The transportation community sees MaaS as a potential major game changer for mobility services (Li and Voege 2017). MaaS is based on e integrating various forms of transportation services into a single mobility service, accessible on demand by individual customers. To meet a customer request, a MaaS operator facilitates a diverse range of transportation options, including public transportation, ride-, car- or bike-sharing, taxi, car rental/lease, or a combination thereof. MaaS can offer added value through use of a single app to provide access to various mobility services with a single payment channel, rather than multiple tickets and payments to various operators.

According to Sahala (2017) from Forum Virium Helsinki, the city's innovation agency, MaaS originally was a vision, a change of mind-set, and a new culture of placing the customer in the center of providing mobility services. In Sahala's view, MaaS does not compete with transportation services and will not replace any transportation services. MaaS is a way to access (plan, buy, and use) all transportation services. In addition, MaaS is a new sales channel for various transportation services. A vision of MaaS integrating various transportation services from different operators in Helsinki may include local public transportation operators (HSL and HRT), the local taxi company (Menevä), the regional train operator (VR), car rental companies (24Rent and EKORENT), carsharing company

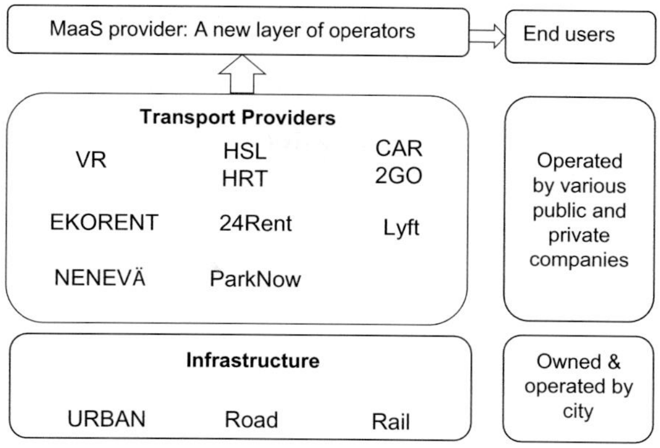

Fig. 12.1 A vision from Helsinki's public authority: MaaS provider as a new layer of mobility operators (Author's elaboration)

(Car2Go), the Transportation Network Company (Lyft), and car park operators (ParkNow). A MaaS provider is a new layer between end users and transportation service operators (Fig. 12.1). The service providers can be owned by local authorities or the privately sector, as either local or international companies. Their data formats, ticket/service sales, and business models vary, presenting great challenges to a MaaS provider.

MaaS should be the best value proposition for an end user by helping meet their mobility needs and solve the inconvenience of planning and booking individual journeys as part of the entire system of mobility services (MaaS Alliance 2017). The principle of MaaS is that a traveler does not need to search for information on individual transportation modes to find a suitable combination of modes and purchase tickets for each one from various operators. MaaS is an ideal integrator of various mobility services to provide maximum convenience of travel experiences without using private cars.

The revolutionary aspect MaaS has brought to the personal mobility market is the concept of a monthly subscription combining various transportation services, which was previously only applicable to public transportation. A monthly or yearly subscription to public transportation (sometimes known as a season ticket) is already widely used in many cities and regions. These tickets offer unlimited use of various public transportation services in a given city or region, such as commuter trains, metro trains, and buses, even though these services may be provided by different operators. Recently, some cities started to include bicycle-sharing in monthly subscriptions. For example, holders of Mobil Dusseldorf season tickets can use local public transportation and shared bicycles for up to 240 min per day (Eryilmaz et al. 2014).

MaaS can offer monthly subscriptions with a larger combination of various transportation services. With a MaaS package, a user can have unlimited usage of public transportation, a certain number of taxi trips, and a certain number of days using car rental, as well as other benefits. Such a business model is inspired by the current personal communication market business plan. Previously, phone bills were based on usage. Mobile phone companies then started to offer a package that included calls and texts. Now, many mobile phone companies offer a package including a certain numbers of calls, unlimited text messages, and a certain amount of usage of the Internet. MaaS simply takes this subscription model into the mobility market (Fluegge 2017).

MaaS has been marketed as a key innovation in the transportation sector that may change current models of transportation provision in urban areas. As a pioneer in MaaS provision, Finland aims to take advantage of the innovation. Notable evidence of this is the recent change in Finland's Transport Code, the Act on Transport Services. The change is a consequence of the emergence of MaaS (Huhtanen 2016). The Act brings changes to the current state of the transportation market that traditionally were strictly regulated and guided by public measures. It aims to promote fair competition in the passenger transportation market and competitiveness of the service providers for both passenger and goods transportation. The Act sets out that essential data on transportation services should be openly available. All transportation providers are required to allow access via open APIs to give information on timetables, routes, ticket prices, and real-time location data (Finnish Government 2017). In addition, Helsinki Regional Transport (HSL) created an open retail platform for single tickets that allows anyone anywhere to purchase single tickets for retail sale (HSL 2017). HSL invites all service provides to develop new mobility services using the interface.

It is difficult to develop the necessary legislation frameworks, as MaaS service deployments are still largely experimental, creating uncertainty about how they will operate, how consumers will use them, and how they will affect society (POLIS 2017). Nevertheless, it is possible to draw conclusions and make first recommendations, based on a detailed analysis of the first deployments of the MaaS approach and the lessons learned.

12.3 Case Studies: Four Examples of Current MaaS Services

Although the principle of MaaS is to provide planning and payment to various transportation services on a one-stop-shop basis (König et al. 2016), MaaS actually comes in different forms. There is no unified form of implementation of MaaS since business models, uses, and technology options vary. Various options may result in various impacts on individual travel behavior and spending, on society and the environment more generally, as well as the local and global economy. Some MaaS providers are simply an extension of local public transportation providers, while others are commercial companies. To demonstrate various versions of MaaS, four services in different countries were selected for an in-depth analysis of the roles of public authorities in MaaS deployment.

These case studies were chosen because they cover the range of commercial setups and operational characteristics for current MaaS deployment, providing a

Table 12.1 Comparison of the four MaaS services

Name	Go Denver	UbiGo	Whim	WienMobil
Provider	Xerox	Go Smart	MaaS Global	Upstream
Provider type	IT company	Research project	Commercial company	Publically owned
Operation city	Denver	Gothenburg	Helsinki	Vienna
Key features	• No payment function • Uses existing back-office	• First trial of MaaS operation • MaaS proof-of-concept	• First commercial MaaS service	• An extension of local public transportation services

balanced, comprehensive analysis of this rather untested approach. The systems analyzed in more detail include a setup allowing integration of shared mobility providers into MaaS (Go Denver), a research project testing the full subscription model (UbiGo), a commercial setup by MaaS Global (Whim), and a system operated by the public sector as an extension of local public transportation (WienMobil). Table 12.1 shows some basic information of the four MaaS services.

These case systems are operated by different types of providers and representing different business models, which are closely linked to roles relevant public authorities have played in their deployments and operation. The analysis allowed for defining enablers and the corresponding roles of public authorities, demonstrating that these roles can vary, based on their policy objectives, technical capacities, regional and national public transportation operations, and relevant regulations.

12.3.1 Go Denver in Denver, USA

Go Denver, powered by Xerox, is an app launched in March 2016. It allows users to plan trips in the Denver metropolitan area using various transportation modes. The app will show available transportation modes, travel times, arrival times, trip costs, calories burned, and carbon emissions (Go Denver 2017). Go Denver does not provide for payment of transportation services, but it is very useful for finding the best transportation option for individual trips.

The Go Denver app covers public transportation, taxi, private transportation (including car, motorbike, or bicycle) platforms, mobility service providers (such as Lyft, Zipcar, car2go, or bi-cycle sharing), and other transportation. Different color codings show which transportation modes are available, based on user profiles and preferences. For example, if the user does not sign up with Car2Go, or the Car2Go app is not installed in the same device as Go Denver, it is not included in trip planning (shown with a gray tab in the app). If the user has an account with Lyft, and the Lyft app is installed on the same device as Go Denver, Lyft is included in trip planning. After the user confirms a planned trip, the Go Denver app will automatically open the Lyft app to book the ride, according to the planned time, origin, and destination.

Since Go Denver does not have a payment function, Xerox could not generate profits from sale of mobility services. An interview with a former director of Xerox confirmed that the company did not receive payment from either the city authority or any transportation providers. Xerox developed the app based on its existing IT infrastructure, as it had provided the back-office facilitate for public transportation services in Denver, including management, information, payment, and clearance. While Xerox used the existing platform for Denver's public transportation service to support app operations, the city authority provided full support to enable Xerox to carry out the service without additional charge. The city authority also played a significant role in promoting the app to encourage locals to use it to support the city's sustainable urban mobility policy, as the app can be downloaded from the city government Web site as a public service. Currently, Xerox aims to generate income by analyzing user data. The city and Xerox are planning to sell the integrated platform to other MaaS providers that can then establish their own business.

12.3.2 UbiGo in Gothenburg, Sweden

UbiGo is an early pioneer of the MaaS concept and was developed by the consortium of a research project named Go: Smart. It was a collaborative project between public authorities, industry, academia, and end users. According to Sochor et al. (2014), "The idea with UbiGo is to offer a combination of transportation modes that can meet individuals' transportation needs." UbiGo integrates different transportation modes and shares resources in a sustainable manner to bridge the gap between private and public transportation by creating a broker for daily travel that offers a tailor-made mobility plan adapted to an individual traveler's needs. By uniting different transportation modes (public transportation; taxi-, car-, and bike-sharing; and car rental), UbiGo offers a simple, flexible, and value-for-money monthly subscription that the entire household can access via a mobile app. UbiGo was the first implementation of the MaaS monthly subscription concept, even though the term MaaS did not yet exist.

More important, the project carried out a six-month trial from autumn 2013 to spring 2014, with participation of 80 households, or approximately 200 individuals. The trial provided an excellent opportunity to evaluate usage of the new service and its effects on users' travel behaviors. Evaluation results from the trial (Sochor et al. 2015) showed that users primarily experienced many practical advantages. For example, it became easier to pay for trips and gain more control over household transportation expenditures. In the final user survey, 93% of participants were satisfied with their trips, partly because the service provided access to more transportation modes and the opportunity to adapt mode choice to each individual trip context, but also because the service created the possibility to try something new and different. One of the most important outcomes of the trial was that many people overestimated how much car access they really needed, and some car-owning participants felt they no longer needed to own their car. Even participants who did not own cars found added value in UbiGo, due to its packaging of services.

The project demonstrated that many actors must cooperate, and many different practical and technical aspects must be integrated to facilitate the implementation of MaaS. To enable a wide uptake of MaaS, there must be deregulation in many aspects of transportation services. In many countries, including Sweden, public transportation services are part of public services and operated by publicly owned companies with government subsidies, so reselling public transportation tickets by private companies is forbidden. To enable UbiGo's operation, in Gothenburg, Västtrafik, the agency responsible for public transportation services temporarily allowed reselling of its tickets. Since 2017, Västtrafik has officially allowed reselling of its tickets to support UbiGo continuity in Gothenburg (Smith et al. 2017).

12.3.3 Whim in Helsinki, Finland

Whim was the first service using the term MaaS, which was provided by a commercial company, MaaS Global. Whim offers pay-as-you-go or monthly subscriptions. Whim offers three options in Helsinki: Whim to Go, Whim Urban, and Whim Unlimited, to meet different demands of mobility services (MaaS Global 2017) (see Table 12.2):

The three packages are targeted at different user groups with different travel demands. Whim to Go is targeted at those who are trying Whim for the first time or simply do not travel that much. Whim Urban is for regular travelers who could occasionally use the flexibility of a taxi or car. Whim Unlimited offers a modern alternative for owning a car. At a price similar to owning a car, users can get unlimited access to public transportation, taxi, or a car, according to their daily needs.

Table 12.2 Various packages provided by Whim in Helsinki

	Whim to Go	Whim Urban	Whim unlimited
Monthly payment	Free	49 €	499 €
Local public transportation	Pay per ride	Unlimited single tickets	Unlimited single tickets
Taxi (5 km radius)	Pay per ride	10 € per ride	Unlimited
Car	Pay per ride	49 € per day	Unlimited

Author's elaboration, adapted from MaaS Global (2017)

Whim Urban covers unlimited local public transportation, and its price (EUR49) is slightly cheaper than the current individual monthly public transportation ticket (EUR54.7) (HSL 2017). In other words, Whim users get a discount on public transportation season tickets. Whim is a new reselling channel for public transportation tickets. However, the Helsinki public authority confirmed that Whim cannot make a profit from reselling public transportation tickets. Whim might need to make profits from taxi companies that expect increased use by Whim users. Although there is no data to confirm the increased use of taxis due to Whim, the Helsinki public authority raised concerns as to whether or not the commercial MaaS service is making urban mobility more sustainable.

MaaS in Helsinki follows the mobile phone contract model of bundling various services. In this case, this business model resulted in significant increase in usage of mobile phones to access the Internet. To address this increased demand, mobile phone operators consequently must increase their capacity by updating their infrastructure at their own cost. However, according to an interview with the city authorities, there are concerns that, following the mobile phone service bundling analogy, users would increasingly use shared cars or taxis included in the MaaS package, resulting in increased travel demand. This is different from mobile phone operators, as MaaS operators do not own the road infrastructure. Therefore, this increased demand must be addressed by public spending. Increasing road capacity is difficult, even impossible in many cities, and it goes against sustainable transportation policy for the past decades. Therefore, the potential negative impacts from MaaS should not be overlooked. Public authorities should provide guidelines and regulations, if necessary, to direct MaaS operators toward sustainable urban mobility goals.

12.3.4 WienMobil by Upstream in Vienna, Austria

In Vienna, Upstream provides a number of MaaS-like services for different types of customers, such as WienMobil for public transportation users and "Jö Bin schon da!" for business users (Stöckl 2017). Upstream is the IT provider of Vienna's public transportation operator, Wiener Linien, which owns 51 percent of Wiener. The other 49% is owned by Wiener Stadtwerke, the public works authority in Vienna, responsible for infrastructure, transportation, and energy. With WienMobil, a user can access public transportation and various other transportation services provided by private operators in Vienna. Such private transportation services include bike-sharing from CityBike and NextBike; carsharing from DriveNow and Car2Go; taxis from two taxi companies; car rental from Europcar; and parking.

In the second phase of WienMobil, which was launched in early 2018, there is a new focus on opening up collected data and service innovation based on the data science and analytics aspect of MaaS that is not covered elsewhere. The service features the Mobility Identity Wallet, based on blockchain technology for single-point, secure, back-office payment processing. In a first step, the legal aspects for implementing such a system will be clarified. Based on the data collected, dynamic traffic management, mobility management, and pricing are being tested.

This service is not aimed at being profitable, but instead improving users' public transportation experience, providing incentive for a shift toward more sustainable modes than private cars, and contributing to the wider public good. Upstream uses the existing transportation integration platform operated by the company, which made accessing data and ticket sales easier. Currently, WienMobil is the only public transportation-led MaaS service. UITP has promoted WienMobil as a success story for public transportation companies to offer MaaS services to their customers (UITP 2017). It created a good example to demonstrate that MaaS can be an extension to existing public transportation, which can be a model to ensure MaaS' contributions to sustainable urban mobility. However, such models may limit competition of MaaS services. However, public authorities may argue that if public transportation services remain a noncompetitive market, then why should MaaS be different?

12.4 Analysis: Current Public Authorities' Practice for MaaS

The case studies described above show that MaaS services, providers, and business models vary widely. Despite the diversity in MaaS service provision, public authorities have played an important role in implementing all four services. Their roles can also vary, depending on local public transportation provision, and social and demographic situations. While some cities see MaaS as a complementary service to their public transportation services, some consider MaaS a sole commercial service evolved from the currently available digital infrastructure.

From a public authority perspective on how to facilitate implementing MaaS services, its role may be primarily categorized into either an access-only or public transportation extension model. An access-only model is a fully market-driven approach, in which a private company develops and operates the MaaS service. A public transportation extension model is a public entity and/or public transportation company that owns or controls the MaaS service.

The public authorities in Helsinki have been using the access-only model to support development of MaaS services. Public authorities in Finland see themselves as a facilitator of any new mobility services which can improve mobility access in their cities. Through the new Act on Transport Services, the Finnish national government has made significant efforts into preparing for the legal framework and digital infrastructure. It also focuses on preparing infrastructure, such as making street parking spaces for carsharing.

WienMobil is a good example of the public transportation extension model, as the MaaS service is provided by a public transportation-owned company based on an existing transportation integration platform. Since this model may potentially reduce competition in MaaS provision, Vienna is considering opening its mobility

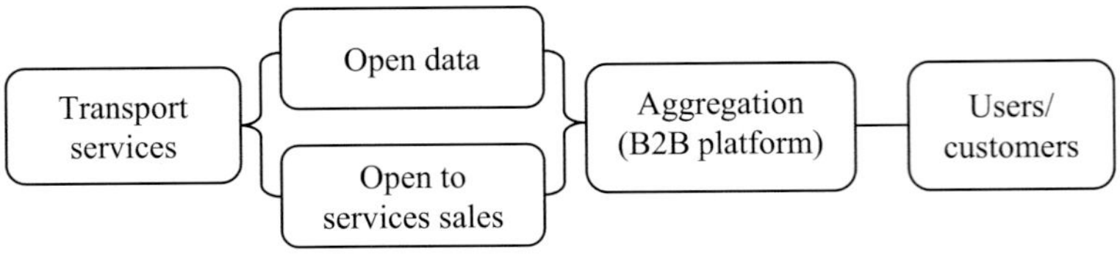

Fig. 12.2 Enabler chain of MaaS (Author's elaboration)

aggregation platform provided by Upstream to allow other private companies access to redistribution of tickets and services under pre-defined conditions. Integrating various services by creating a single platform allows a third party to create a business case. This combines the access-only and public transportation extension model. Under this model, public authorities must invest in developing the MaaS platform. However, the integration platforms are provided by private companies in many cities, such as Xerox in Denver. In this case, investment in the MaaS platform may come from the private sector if it sees this as a viable business opportunity.

Despite the various forms of MaaS, its enabling implementation chain has the following minimum components: transportation services, such as metro, buses, taxis, carsharing, and car rental; services that operate without MaaS but may have their form and market influenced by MaaS in the future; open access to data (APIs, interfaces) about their transportation services, such as static information (routes and timetables), and real-time information (location and availability of vehicle); access to ticket and services sales and agreements on resale; aggregation to integrate available options and payments; and users/customers, who will compare MaaS with existing transportation provision and decide if they will subscribe. Figure 12.2 shows the enabler chain.

To support MaaS implementation, public policy and spending should focus on further regulating open data and open API for ticket sales and payment, supporting negotiation of service agreements, building an integrated, business-to-business (B2B) platform, and promoting MaaS to users.

12.4.1 Further Regulation on Open Data and Open API to Ticket Sale and Payment

While the open-data policy at national and EU levels has made public transportation data available to developers, there is a lack of policies to encourage or require private transportation operators to open their data. In addition, there is a need for open ticket (or service) sales, such as open APIs to sell tickets, from public transportation and private service providers. Public authorities can focus on removing boundaries to support new mobility innovation and services, such as promoting digitalization in transportation by ensuring availability of digital, open, interoperable data on the transportation market, as enabled through the Finland Transport Act. Adequate legislative and legal conditions to enable any company to provide MaaS services to ensure fair competition can lead to savings in publicly funded passenger transportation services, better services for customers, and new business opportunities for companies. To facilitate a wide uptake of MaaS, there must be deregulation of many aspects of transportation services, particularly in public transportation ticketing services. The case in Gothenburg shows that deregulation may be done at a city, rather than a national level, as in Finnish. However, if current practices at city and national levels show positive impacts on urban mobility, environment, and economy, the deregulation can be moved to the EU level.

12.4.2 Supporting Negotiation of Service Agreements

The Denver government supported the implementation of Go Denver by facilitating cooperation between the service provider and various transportation operators. The city's goal in supporting MaaS is to give local residents more option by ideally making travel in the city cheaper and more convenient. Similar to Denver, Transport for West Midlands, UK, the public authority responsible for transportation services in the region, supported implementing Whim by facilitating cooperation between the MaaS provider and local transportation providers, but without direct financial support. Transport for West Midlands commissioned a university to carry out a small research project to study users' behaviors and the effects of MaaS in the region after launching the Whim service. Such results, once available, will provide an important reference to the public sector for future policy development.

12.4.3 Building an Integrated Platform (B2B Platform)

Building an integrated platform that can be used by a MaaS provider (B2B platform) can be a direct investment from the government or private companies. Since many cities have integrated platforms for public transportation services, it may not be expensive to extend the platform to more transportation modes. If the city's platform is provided by private companies, the city can require the B2B service in its public procurement. In a B2B MaaS platform, data from various transportation service providers are collected into a single platform layer, enhanced with features, such as multimodal journey planning or mobile payment services, and are made available to third parties. However, such a B2B platform must address quality of data and its maintenance, such as how to integrate a new mode when it emerges.

12.4.4 Promoting MaaS to Users

Public authorities, particularly at the local level, can play an important role in promoting MaaS to their citizens by disseminating information about its benefits. However, public authorities have the power to significantly scale up the MaaS market if the mobility service is connected to city planning and land use. In a recent feasibility study of a MaaS service in Malmö, Sweden, the service was proposed to work with property developers who would offer a MaaS package to their residences in new properties that have reduced parking facilities (Indebetou et al. 2016). However, this would not be a workable model in cities that already have advanced sustainable urban mobility. According to an interview with the transportation department in Copenhagen, although three public transportation operators are interested in MaaS in principle, the idea of connecting MaaS to accommodation is less feasible in Copenhagen than elsewhere, due to a lack of the necessary preconditions for parking requirements, since the city has applied a policy for reducing private car use for many years.

12.5 Recommendations: Public Authorities' role for MaaS

While many public authorities see MaaS as a clear trend going forward, implementing it has rarely been included in local or national transportation policies, even though demonstration, research, and operation of various types and formats of MaaS are implemented in cities in both Europe and the USA. Technologies, business models, and effects are still in flux. There is a lack of research into either potential implementation scenarios or the associated effects. There are currently only a few best practices to share from cities in which MaaS has been in operation. Much of the publicity on MaaS originates from providers, which may not be evidence-based but driven by commercial interests. Therefore, public authorities are not well-informed, leading to difficulties in formulating adequate policy frameworks. With the paradigm shift in transportation regulations in Finland, many public authorities feel the pressure of being obligated to support the implementation of MaaS without evaluating potential effects.

If a public authority is interested in implementing MaaS in its area, a series of questions would be raised, such as what kind of investments are necessary, how to procure a system, what to procure, and how to evaluate the benefits to citizens and society as a whole. Creating new markets or promoting certain technologies should only be supported if they contribute to transportation policy objectives. A balanced governance model with public sector leadership should be sought to ensure an equitable, sustainable transportation system. Currently, KPIs to evaluate the mobility service and its contributions to the environment, energy, and social inclusion have not been set and are not even being sufficiently discussed.

In many countries, public transportation is part of public services provided by local authorities and heavily subsidized by public spending. Governments see the emergence of new services as an opportunity to reduce public spending on subsidies to public transportation services, which is one of the goals of the Finnish Transport Code. The Finnish Ministry of Transport indicates that MaaS can provide an excellent solution for making mobility services more convenient at a cheaper price for both users and society, as MaaS may fundamentally change the definition of public transportation. This may lead to public authorities considering their funding schemes and procurement for public transportation services. There are cases in the USA in which public authorities subsidize shared mobility instead of commissioning public transportation on routes where there is not sufficient demand (TCRP 2016). Public authorities also subsidize shared mobility instead of building new infrastructure to meet increasing demand. The economic factors associated with MaaS should not be ignored by public authorities.

Overall, public authorities must be conscious of their policies, particularly regarding financial and legislative support. The SPICE project to study public financing of innovative and sustainable transportation compiled a list of policy items for public authorities to evaluate whether or not a MaaS service should be supported (SPICE 2017). Examples of such evaluation indicators include KPIs in mobility services, the environment, and energy consumption, and effects on government spending and short-term (such as job creation) and longer-term, local economic development (such as innovation and competitiveness).

Since MaaS is still in its infancy, there is a lack of robust data on its performance and associated societal effects. With MaaS rolling out in many cities, it is feasible to collect data for monitoring its performance and assessing its effects. However, there is a need for an evaluation framework to provide a methodology on data collection and analysis. This should be a priority for transportation research. Public authorities should envisage challenges and opportunities to urban mobility brought by MaaS and be prepared with more data-driven, evidence-based policies and regulations.

References

American Public Transportation Association (2016) Shared mobility and the transformation of public transit. https://www.apta.com/resources/reportsandpublications/Documents/APTA-Shared-Mobility.pdf. Accessed 2 Dec 2017

Banister D (2008) The sustainable mobility paradigm. Transp Policy 15(2):73–80

Davison LJ, Knowles RD (2005) Bus quality partnerships, modal shift and traffic decon-gestion. J Transp Geogr 14(3):177–194

Department of Transport UK (2013) Door to door: a strategy for improving sustainable transport integration

Dotter F (2016) Integrated ticketing and fare policy for public transport. CIVITAS Insight 12

EC (2009) Action plan on urban mobility. COM (2009) 490 Final. 30 Sept 2009

EC (2017a) European urban mobility policy context. https://doi.org/10.2832/827766

EC (2017b) Commission delegated regulation (EU): supplementing directive 2010/40/EU of the European parliament and of the council with regard to the provision of EU-wide multimodal travel information service
Eryilmaz E, Kagerbauer M, Schuster T, Wolf O (2014) Collaborative management of intermodal mobility. IFIP Adv Inf Commun Technol 434:713–721
European Platform on Mobility Management (EPOMM) (2017) The role of Mobility as a Service on mobility management
Finnish Government (2017) Good and flexible transport services through a new act. Press Re-lease, 25 May 2017
Fluegge B (2017) A new service paradigm. In: Smart mobility-connecting everyone. Springer
Go Denver (2017) http://godenverApp.com/. Accessed 2 Dec 2017
Goodall W, Fishman TD, Bornstein J, Bonthron B (2017) The rise of Mobility as a Service. Deloitte Review
Goodwin PB (1993) Car ownership and public transport use: revisiting the interaction. Transportation 20:21–33
Hall J, Palsson C, Price J (2017) Is Uber a substitute or complement for public transit? Working paper. University of Toronto
House of Parliament UK (2014) Big and open data in transport. POST-PN-472
HSL (2017). HSL to launch world's first public transport retail interface open to everyone. Accessed 7 Dec 2017
Huhtanen S (2016) Legislation, an enabler for MaaS: finnish Transport Code. In: Presented at the 7th IRU international taxi forum, Cologne, 5 Nov 2016
Iacobucci J, Hovenkotter K, Anbinder J (2017) Transit systems and the impacts of shared mobility. In: Meyer G, Shaheen S (eds) Disrupting mobility, pp 65–76. Springer, Berlin
Indebetou L, Kerttu J, Milton J, Wendle B (2016) Feasibility study MaaS–a business case for EC2B. Project report from Climate-KIC. 31 Oct 2016
ITF-OECD (2016) Shared mobility: innovation for livable cities. OECD, Paris
Kamargianni M, Matyas M, Li W, Schäfer A (2015) Study for "Mobility as a Service" concept in London. UCL Energy Institute
König D, Eckhardt J, Aapaoja A et al (2016) Deliverable 3: business and operator models for MaaS. MAASiFiE project funded by CEDR
Li Y, Kristensen JP (2014) Transport data marketplace: an analysis of user requirements. Paper presented at the 10th ITS European congress, Helsinki, 16–19 June 2014
Li Y, Voege T (2017) Mobility as a Service (MaaS): challenges of implementation and pol-icy required. J Transp Technol 7:95–106
MaaS Alliance (2017) White paper on guidelines and recommendations to create the foundations for a thriving MaaS ecosystems. Brussels
MaaS Global (2017) https://whimApp.com/ Accessed 2 Dec 2017
Midgley P (2011) Bicycle-sharing schemes: enhancing sustainable mobility in urban areas. Background paper at UN Department of Economic and Social Affairs Commission on Sustainable Development 19th Session, New York, 2–13 May 2011
POLIS (2017). Mobility as a Service: implementations for urban and regional transport. Discussion paper, 4 Sept 2017
Sahala S (2017) How to procure MaaS. Presented at the SPICE project webinar, 27 Sept 2017
Shaheen S, Guzman S (2011) Worldwide bikesharing. UC Berkley Access Mag 1(39)
Smith G, Sochor J, Karlsson M (2017) Mobility as a Service: implications for future mainstream public transport. Paper presented at the 15th international conference series on competition and ownership in land passenger transport. Stockholm, 13–17 Aug 2017
Sochor J, Strömberg H, Karlsson M (2014) Traveller's motives for adopting a new, innovative travel service: Insights from the UbiGo field operational test in Gothenburg, Sweden. Paper presented at the 21st world congress on intelligent transport systems. Detroit, 7–11 Sept 2014
Sochor J, Strömberg H, Karlsson M (2015) Implementing Mobility as a Service: challenges in integrating user, commercial, and societal perspectives. Transp Res Rec 2036:1–9

SPICE (2017) Report from workshop on future roles of public authorities in Mobility as a Service (MaaS). Vienna, 10 Apr 2017

Stöckl G (2017) Upstream–next level mobility. Presented at SPICE workshop on future roles of public authorities in Mobility as a Service (MaaS). Vienna, 10 Apr 2017

TCRP (2016) Research Report 188 on shared mobility and the transformation of public. TRB Publication. https://doi.org/10.17226/23578

Uber (2016) Making ridesharing more affordable in Pinellas County. https://www.uber.com/blog/tampa-bay/psta-td/. Accessed 2 Dec 2017

UITP (2011) Becoming a real mobility provider combined mobility: public transport in synergy with other modes like car-sharing, taxi and cycling. UITP position paper

UITP (2017) The Mobility as a Service (MaaS) success story: WIENMOBIL. http://www.uitp.org/The-Mobility-as-a-Service-MaaS-success-story-WienMobil. Accessed 23 Dec 2017

Wharton (2017) Demographic shifts: shaping the future of car ownership. http://knowledge.wharton.upenn.edu/article/demographic-shifts-shaping-future-car-ownership/. Accessed 2 Dec 2017

Yanying Li holds a Ph.D. in transportation planning and engineering from the University of Southampton, UK, and a master's degree in civil engineering from Tsinghua University, China. She currently works in the European Climate Foundation as a City Liaison Officer to assist cities around the world to reduce greenhouse gas (GHG) emissions. From 2008 to 2017, she worked as Senior Manager at ERTICO–ITS Europe, where she led a number of EC-funded international cooperation projects working with cities from Europe, East Asia, the Middle East, and Latin America on the implementation of sustainable urban mobility policies and technologies. Prior to joining ERTICO, she worked as a Research Fellow at the University of Southampton.

Part V
Conclusion

Chapter 13
Conclusion

Maxime Audouin and Matthias Finger

Technology is the answer. But what was the question?
Eric Price (1966)

Abstract In this concluding chapter, we summarize the three key insights drawn from the different contributions of this book. We then reflect on our notion of smart transport by highlighting that, although it might be a necessary condition, it might unfortunately not be sufficient to go toward the post-car system. Thus, we propose leads for more research to be conducted in order to ensure that smart mobility solutions are not only "smart," but are also aligned with a sustainability paradigm. In particular, we highlight the need to conduct empirical research on the impacts of existing smart mobility solutions so as to eventually better understand if those are able to keep their promises in terms of sustainability, which should ultimately enable researchers to establish the missing link between the governance of smart transportation systems and the impacts of such governance approaches on the overall performance of transportation systems.

Keywords Smart and sustainable mobility · Paradigms alignment
Framing · Multi-level governance

13.1 Introduction

By looking specifically through a series of case studies, at the institutional mechanisms through which smart transport solutions have developed, this book has aimed to offer a comprehensive overview of the various research endeavors that focus on the governance of smart transportation systems. It has conceptualized smart transportation systems as transportation systems comprised of one or more of the following smart transport "pillars": shared, automated, electric, or integrated mobility solutions. In this concluding chapter, we present the three main insights

M. Audouin (✉) · M. Finger
Chair Management of Network Industries (MIR), College Management of Technology (CDM), EPFL, Lausanne, Switzerland
e-mail: maxime.audouin@epfl.ch

M. Finger
e-mail: matthias.finger@epfl.ch

M. Finger and M. Audouin (eds.), *The Governance of Smart Transportation Systems*, The Urban Book Series, https://doi.org/10.1007/978-3-319-96526-0_13

that we have taken from the different contributions exposed in this book. The first insight is that the current regulatory responses of governments to the development of smart transportation solutions deals are largely inappropriate. Public authorities need to urgently define new regulatory answers to better integrate new transport solutions to their existing transportation systems. The second insight is the confirmation of the assumption that the way governance takes place is basically shaped by how problems are framed. The third and final insight is the increasingly importance of the local level in the development of innovative mobility solutions, and consequently the need to consider that governance should be multilevel to guarantee a successful development of smart transportation systems.

13.2 The Inappropriateness of Current Regulatory Response to Smart Transportation Solutions

As we have seen in the book, most smart transport solutions are actually being proposed by the private sector, usually overcoming the capacity of the public sector to react. As an answer, the public sector usually develops regulations for smart transport solutions based on traditional regulatory approaches. For example, Puche (Chap. 3) showed that public authorities in Latin America regulated TNCs using regulatory frameworks building on traditional taxi regulation, but that such approaches were ineffective, as smart transport solutions providers often found ways to circumvent the proposed regulations, claiming their foundational differences with incumbent actors. Similarly, Voege (Chap. 5) showed that most of the regulations being developed for SDVs actually pertained to the safety of automated vehicles and that not enough regulation was being developed to access the operations data of shared automated vehicles. Where public authorities appear to have the opportunity to take a step ahead and kill two birds with one stone by developing data-driven regulations, they appear to have failed in doing so, either because of a lack of vision or because of the slowness of the administrative apparel. Ultimately, Montero (Chap. 2) proposed a new way of looking at shared mobility services providers, paving the way for the development of a new regulatory approach, which could be a solution to the successful inclusion of shared mobility services in urban transportation systems.

13.3 The Way Governance Happens Depends on How Problems Are Framed

According to Dowling (2018: 51), "*governance solutions are influenced by how problems are framed.*" In other words, and applied to transport, the way smart transportation solutions develop actually depends largely on the narratives associated with those and the reasons given to promote their uptake. Throughout this book, different chapters seem to have confirmed this stance. For example, Ryghaug and Skøsvold (Chap. 8) showed that the way in which electric mobility developed in Norway was clearly dependent on the reason why public authorities wanted it to takeoff. Electric mobility was not originally part of an environmentally friendly narrative, but clearly part of an economic development at the country level. This emphasizes the importance of choosing the "right" narrative for a smart transport solution to develop. The choice of developing a DRT service in Nijmegen as a way of reducing both carbon emission and the use of fossil fuel can also be seen as having had an impact on the way it developed (Sharmeen and Meurs, Chap. 12). Faivre d'Arcier and Lecler (Chap. 4) found similar results when looking at the development of carsharing in France and Japan. In the case of France, carsharing was developed by public authorities in order to specifically tackle private motorized travel modal share, resulting in its development as a subsidized service. In Japan, by contrast, the framing of carsharing services as commercial activities resulted in different outputs in terms of usage by citizens and involvement for public authorities. Similarly, Mladenović (Chap. 6) showed that the future of SDVs will basically depend on how their developments are framed and that, in order to frame things correctly, a phase of participatory expansion of the technological horizons of desirable futures needs to be developed.

13.4 Need to Develop a Multilevel Approach to the Governance of Smart Transportation Systems

Last but not least, an element that has emerged from most of the chapters pertains to the need to consider different jurisdictional and geographical levels when studying the governance of smart transport solutions. In most of the chapters that look at past cases of smart transport developments, the authors have indeed emphasized the importance of coordination mechanisms between actors operating at different jurisdictional levels. For example, Smith et al. (Chap. 9) and Li (Chap. 11) showcased that successful development of MaaS schemes was actually dependent on the coordination of actions between governing actors at the national, regional, and local levels. Audouin and Finger (Chap. 10) came to the same conclusion regarding the successful unfolding of smart ticketing schemes and added that actors lying at the supra-national level (EU level, for example) also had roles to play in the development of smart transport solutions. Looking at electric mobility, and more particularly at the Netherlands and Brazil, Rietmann and Lieven (Chap. 7) also showed that the development of incentives and regulations by public bodies at the local, regional, and national levels was determinant for electric vehicles to take off. Ryghaug and Skøsvold (Chap. 8) showed that it was crucial to consider both scales (local and national) in order to have a full understanding of how EVs developed in Norway. In order to better consider the different territorial and jurisdictional levels involved in the development of smart transport solutions, we argue that other approaches, such as the multilevel governance (MLG) framework, might be useful. The MLG is indeed acknowledged as providing researchers with a robust analytical tool to look at "*'arrangement' of policy-making activity performed within and across politico-administrative institutions located at different territorial levels*" (Stephenson 2013: 817). MLG has already been used to look at the governance of sustainable transport policy (Marsden and Rye 2010), but, to our knowledge, has not been widely used to look at the development of smart transport solutions. By bringing territorial and jurisdictional dimensions into the analysis, the MLG can be understood as being complementary to the Multi-level Perspective (MLP) on sustainability transitions framework, which is often criticized for lacking a geographical dimension (Coenen et al. 2012). While attempts to conciliate both frameworks have been limited (see for example Hoffmann et al. 2017), academics have been calling for the development of such approaches for quite some time (Whitmarsh 2012).

Having summarized the three main learnings that one can withdraw from the contributions gathered in this book, we will now offer some recommendations for future research related to the governance of smart transportation systems.

13.5 The Way Forward

This book was built on the assumption that smart transportation systems will help pave the way out of the incumbent automobility regime. But there is actually an urgent need to validate (or refute) this hypothesis through empirical research. Indeed, a growing body of literature has criticized smart transport solutions for not being able to keep their promises in terms of sustainability and for potentially not improving transport conditions compared to the status quo. According to Docherty et al. (2017), private actors proposing smart mobility services are, because of the business models adopted, actually more interested in creating a market where there is more and not less mobility. Consequently, if smart transportation systems unfold without any public intervention, there is a real risk that transportation systems will function worse than they do today (Currie 2018; Hensher 2018). In a similar fashion, Ryghaug and Skøsvold (Chap. 8) warned of some unclear aspects of road transport electrification impacts in Norway. Although the increasing share of EVs in the total number of vehicles in circulation on Norway's road might be beneficial from an environmental perspective (less greenhouse gases emitted), the possibility of entering in the future an "electric-automobility system" must not be ignored. Puche (Chap. 3) also questioned the real ability of e-hailing services to significantly tackle urban congestion due to their low-occupancy capacity and the fact that they often circulate "empty" of any passengers from the end of one ride to the beginning of the next one. Smith et al. (Chap. 9) and Li (Chap. 12) also highlighted the possibility of having unsustainable MaaS schemes come to life, relying more on low-occupancy vehicle than on public transit. Similarly, Audouin and Finger (Chap. 10) questioned the capacity of a smart ticketing solution to, by itself, induce modal shift and impact private vehicle ownership. To understand the real impact that new transport solutions (branded "smart" in this book) are producing on existing transport systems, we argue that much more empirical research needs to be conducted on such solutions. For example, it is currently unclear what impact smart transportation solutions might have on existing transport systems if taken separately. For example, if all vehicles were to become electric tomorrow, we would tackle an important part of the pollution problem linked to road transport. But the system would not be more integrated, and there would probably be as many people owning cars as is the case today (which would mean a limited impact on congestion, for example), which would have to be human-driven (hence having a limited impact on road safety). Similarly, if all cars were to become automated tomorrow, this would probably have an impact on the efficiency of road transport and perhaps on road safety, but it is unclear what the impact on the environment would be. Indeed, if the only difference compared to today would be to have cars driving themselves, but still running on fossil fuel, still owned by individuals and still part of an un-integrated transport system, one might be concerned about the energy that would be required to allow those to circulate. One could also ask the same questions with all vehicles becoming shared, but still being driven by individuals, powered by fossil fuel and not being integrated with other transport solutions (such as public transit), and with the transportation system becoming completely integrated, but with vehicles not being more shared than today, running on fossil fuel and being driven by humans. Although there is limited data available to show the impact of smart transport solutions (due to the very young age of those solutions), there are increasing cases developing around the world, which should allow for the development of more evidences related to the impact of such solutions on cities, which should ultimately allow researchers to link governance structures with performance of (smart) transportation systems.

In order for smart transport solutions to become sustainable (Lyons 2016), there is also a need to develop more travel demand management schemes. It is only by managing the travel demand that public authorities might prevent it from exploding as a result of the introduction of smart transport solutions. In particular, the use of measures such as congestion charging schemes or transit-oriented development still appear relevant to guarantee a sustainable future for transportation systems. Those might be categorized as transport policies, which are necessary according to Urry (2004), along with new propulsion technologies, smart cards, new consumption patterns, and the ICTs, to go toward the post-car system. Transport demand management approaches might not be as "smart" as the different transport solutions presented in this book, but they are probably less likely to produce unanticipated consequences than technology-driven (smart) transport options unfolding nowadays. Thus, we also believe that more research focusing on the efficiency of dedicated schemes to manage travel demand must be conducted in order to give public authorities all the cards they need to enhance the transition towards sustainable transport systems.

Although things might be a little bit more advanced in the area of electric mobility, smart transport solutions are still in their infancy. Thus, their uptake depends on how individuals react and receive them. While we have seen that, from a sustainability perspective, it would be far better for SDVs to develop as shared vehicles (rather than as personally owned ones), it is currently far from clear whether this will be the case, or if SDVs will be owned as much as conventional human-driven cars are nowadays. Similarly, it is unclear what impact integrated mobility schemes, such as MaaS, will have on people's travel behaviors. According to Pangbourne et al. (2018), using taxi-like services might become so cheap and convenient in the future that there is a real rebound effect threat associated to MaaS. We believe that research must be conducted to predict those rebound effects, either using agent-based modeling [as done by the ITF (2017)] or randomized control experiments. Similarly, research on existing cases must be conducted to see how individuals change their behaviors when they embrace a smart transport solution. In particular, the use of practice theory [see, e.g., Watson (2012)] might be relevant to do so.

References

Coenen L, Benneworth P, Truffer B (2012) Toward a spatial perspective on sustainability transitions. Res Policy 41(6):968–979

Currie G (2018) Lies, damned lies, AVs, shared mobility, and urban transit futures. J Public Transp 21(1):19–30

Docherty I, Marsden G, Anable J (2017) The governance of smart mobility. Transp Res Part A: Policy Pract. http://dx.doi.org/10.1016/j.tra.2017.09.012

Dowling R (2018) Smart mobility: disrupting transport governance? In: Marsden G, Reardon L (eds) Governance of the smart mobility transition, pp. 51–64. Emerald Points

Hensher D (2018) Tackling road congestion—what might it look like in the future under a collaborative and connected mobility model? Transp Policy. https://doi.org/10.1016/j.tranpol.2018.02.007

Hoffmann S, Weyer J, Longen J (2017) Discontinuation of the automobility regime? An integrated approach to multi-level governance. Transp Res Part A: Policy Pract 103:391–408

International Transport Forum (2017) Shared mobility—innovation for liveable cities. ITF and OECD, Paris

Lyons G (2016) Getting smart about urban mobility—aligning the paradigms of smart and sustainable. Transp Res. Part A. http://dx.doi.org/10.1016/j.tra.2016.12.001

Marsden G, Rye T (2010) The governance of transport and climate change. J Transp Geogr 18 (6):669–678

Pangbourne K, Stead D, Mladenović M, Milakis D (2018) The case of Mobility as a Service: a critical reflection on challenges for urban transport and mobility governance. In: Marsden G, Reardon L (eds) Governance of the smart mobility transition, pp 33–48. Emerald Points

Stephenson P (2013) Twenty years of multi-level governance: where does it come from? What is it? Where is it going? J Eur Public Policy 20(6):817–837

Urry J (2004) The "System" of automobility. Theor Cult Soc 21(4/5):25–39

Whitmarsh L (2012) How useful is the multi-level perspective for transport and sustainability research? J Transp Geogr 24:483–487

Watson M (2012) How theories of practice can inform transition to a decarbonised transport system. J Transp Geogr 24:488–496

Maxime Audouin holds a Bachelor of Science in environmental engineering, and a Master of Science in energy management, both from Ecole Polytechnique Federale de Lausanne (EPFL), Switzerland. Since 2015, he has been enrolled as a Ph.D. student in the Swiss Post Chair in Management of Network Industries (MIR) at EPFL, under the supervision of Prof. Matthias Finger.

His research interest lies in the area of urban transportation systems, urban governance, transportation policy and regulation, as well as technology and innovation management. In his dissertation, he looked at the role that public authorities play in the development of ICT-supported integrated mobility schemes.

Matthias Finger received his Ph.D. degrees in political science and adult education from the University of Geneva, Switzerland. He is known for his expertise in matters of regulation and governance of network industries (postal services, telecommunications, electricity, railways, air transportation, urban public transportation, water). He has been an Assistant Professor at Syracuse University (New York), an Associate Professor at Columbia University (New York), and a Full Professor at the Swiss Federal Institute of Public Administration (Lausanne). Since 2002, he has held the Swiss Post Chair in Management of Network Industries (MIR) at Ecole Polytechnique Federale de Lausanne (EPFL), Lausanne, Switzerland, and since 2010 has directed the Florence School of Regulation's Transport Area at the European University Institute in Florence, Italy.

Glossary

Automobility refers to the dominant mobility regime, which is built around fossil fuel-powered privately owned motorized vehicles and is supported by specific institutions, infrastructures, technology, scientific knowledge, policies, and industries. It is often depicted as a path-dependent nonlinear system that is responsible, among other things, for congestion and air pollution, as well as intensively contributing to climate change.

Battery electric vehicles (BEVs) use electric motors instead of traditional internal combustion engines (ICE) for propulsion and use chemical energy stored in dedicated battery packs to run their electric motors.

Big Data can be understood as high-volume, high-velocity, and high-variety information assets that demand cost-effective, innovative forms of information processing, enabling enhanced insight, decision making, and process automation.

Carpooling refers to the sharing of a vehicle journey by the vehicle driver, so that other people can join the ride, which ultimately enables the driver to share costs of his trip. Carpooling services are usually organized via dedicated online platforms such as BlaBlaCar, where users can search for a trip matching their needs and book and pay for the service online.

Carsharing (referred to as car clubs in the UK) refers to the sharing of a vehicle (or pool of vehicles) between individuals. Carsharing can actually take four different forms. It can either be organized as round-trip carsharing, where carsharing vehicle users must return the vehicle where they picked it up; one-way carsharing, where carsharing vehicle users are able to drop the vehicle used to a different station than the one they picked it from; P2P carsharing, where an individual decides to share his or her own vehicle with other people; or fractional ownership, where several individuals decide to jointly buy or lease a vehicle. Carsharing services are usually organized by dedicated companies that propose digital platforms enabling users to search for a car, book, and pay online.

M. Finger and M. Audouin (eds.), *The Governance of Smart Transportation Systems*,
The Urban Book Series, https://doi.org/10.1007/978-3-319-96526-0

Contactless payment cards (CPCs) are payment cards (credit or debit cards) that are equipped with a microprocessor and internal memory enabling communication between the cards and smart card readers through radio frequency technology. Those are being increasingly welcomed as a new means of transport ticketing.

Demand-responsive transit (DRT) refers to demand-based public transit system. As opposed to having fixed transit lines operating on fixed routes on a fixed schedule, DRT operates when and where users demand it, which they usually do through an application via their smartphones or Web interface. DRT vehicles are usually minibuses that take about 6–14 seated passengers whose routes are determined by a ride-pooling algorithm that finds the shortest route to access the different passengers' destinations. Famous examples of DRT solutions are Chariot (a subsidiary of Ford) and MOIA (a subsidiary of Volkswagen).

Digitalization can simply be understood as the process by which information is transformed to become easily read by computers. In the field of transportation, digitalization is acknowledged as having had a disruptive effect and having enabled new services and business models to come to life.

E-hailing refers to the mobility services provided by transportation network companies (TNCs) that are usually bookable by individuals via a digital platform, supported by a dedicated application on smartphone or via Web interface. In this book, the notion of e-hailing encompasses the other terms usually used in the literature, of ride-sourcing, ride-booking, and ride-sharing.

Information and communication technologies (ICTs) refer to the combination of hardware and software infrastructures that enable data gathering and exchange via specific channels.

Mobility-as-a-Service (MaaS) refers to a digitally supported distribution model that bundles several transport options together and enables users to plan their trips, select the transport option that best suits their needs, and finally book and pay for it via a unique point of access, whether it is an application or a Web site.

Multi-level Perspective (MLP) is a conceptual framework used as a heuristic tool to look at long-term socio-technical transitions. It basically analyses the diffusion of system innovations as a result of the interplay between three key structuring layers: the niche, regime, and landscape layers.

Multimodal transport designates the combination of two or more different transport services, public or private, to accomplish a journey.

Multisided markets can be understood as markets with two or more user groups that provide each other with network benefits. The company facilitating the establishment of a relation between the different users groups of the market is usually called a platform. Multisided markets have existed for a long time, but have really taken off recently thanks to digitalization. Examples of digital platforms in transportation are Uber or BlaBlaCar, which basically provide a digital platform that enables drivers to propose their services and connect with people with mobility needs, that is, potential passengers (and thus customers).

Pay as you go (PAYG) relates to a tariff structure used by some transport ticketing schemes. Under this structure, users basically pay for the amount of mobility services they "consume". Such schemes differ from monthly subscriptions, which are usually charged upfront and often give users unlimited access to mobility services.

Hybrid electric vehicles (HEVs) are equipped with both a conventional internal combustion engine (ICE) and an electric propulsion system with the aim of achieving better performance in terms of fuel consumption than conventional ICE vehicles. A subcategory of HEVs, known as plug-in hybrid electric vehicles (PHEVs), can receive an external electric power source to charge their electric batteries. In this case, the combustion engine acts as a backup when electric batteries are depleted.

Self-driving vehicles (SDVs) refer to vehicles that are able to sense their environment through dedicated sensors, analyze the gathered data, and make decisions accordingly in order to navigate with reduced or without human input. Different levels of vehicle automation exist, ranging from limited assistance to full automation. In this book, the term SDV encompasses the other terms usually employed in the literature, such as automated, autonomous, and driverless vehicles.

Smart Cards are plastic pocket-size cards, with a chip embedded, that can store and, in most cases, process data through radio frequency technology. They are used in transportation as a means of ticketing and often to support integrated ticketing and fare systems.

Smart Transportation Systems are used in this book to designate the ensemble of transport solutions that rely on the use of ICTs. Smart transportation systems can actually be seen as composed of four interrelated categories: shared, integrated, automated, and electric mobility solutions.

Socio-technical systems (STS) are systems in which technological and social components interact with one another, within a given environment. Within such systems, people might influence the development of technological components, which in turn might influence people's behavior and vice versa.

System innovations can be understood as profound transformations in the way societal functions are usually carried out. For example, the automobile can be understood as a system innovation given that, when introduced, it deeply transformed people's mobility and travel habits.

Systems of innovation (SI) (not to be mistaken with system innovations) is a school of thoughts aimed at understanding the economic, social, political, and organizational factors that determine the development and diffusion and uptake of innovations. There are different approaches in the SI approach, ranging from national (NIS) and regional (RSI) to sectoral (SSI) or technological innovation systems (TIS).

Transition management (TM) can be used as a research model or a policy tool. It can be understood as a conceptual framework to study the development of potential system innovations and, more specifically, the governance of systemic transformations of socio-technical systems, usually with sustainability as the overarching goal.

Transportation network companies (TNCs) are enterprises that provide a dedicated digital platform that enables the pairing of passengers with drivers and thus supports the provision of so-called e-hailing services (sometimes referred to as ride-booking, ride-sharing, ride-sourcing, etc.). Examples of famous TNCs are Uber and Lyft.

MIX
Papier aus verantwortungsvollen Quellen
Paper from responsible sources
FSC® C105338

If you have any concerns about our products,
you can contact us on
ProductSafety@springernature.com

In case Publisher is established outside the EU,
the EU authorized representative is:
Springer Nature Customer Service Center GmbH
Europaplatz 3, 69115 Heidelberg, Germany

Printed by Libri Plureos GmbH
in Hamburg, Germany